W9-DEV-905

General Class

FCC License Preparation for Element 3 General Class Theory

by
Gordon West
WB6NOA

Seventh Edition

Master Publishing, Inc.

Gordon West, WB6NOA
FCC Amateur & Commercial Radio License Preparation Study Materials

Technician Class, Element 2
Technician Class study manual
Audio Theory Course on CD
Technician Class book with
    W5YI HamStudy Software

General Class, Element 3
General Class Study Manual
Audio Theory Course on CD
General Class book with
    W5YI HamStudy Software

Extra Class, Element 4
Extra Class Study Manual
Audio Theory Course on CD
Extra Class book with
    W5YI HamStudy Software

GROL-Plus
**General Radiotelephone Operator License
Plus Ship Radar Endorsement**
(with Fred Maia, W5YI)
FCC Commercial Radio License Preparation for
Commercial Elements1, 3, and 8 Question Pools
GROL-Plus book with
    W5YI study software

This book was developed and published by:
Master Publishing, Inc.
Niles, Illinois

Editing by:
Pete Trotter, KB9SMG

Photograph Credit:
All photographs that do not have a source identification are either from the author,
or Master Publishing, Inc. originals.
Cover photo by Larry Mulvehill, WB2ZPI, *CQ Magazine.*

Cartoons by: *CD mastering by:*
Carson Haring, AC0BU   Mike Koegel, N6PTO, Studio 49

*Thanks to the following for their assistance with this book: Suzy West, N6GLF;
Shorty & Susan Stouffer, K6JSI & K6SLS; Ed Collins, N8NUY; Howard Herhold,
KC4ZYC; Bill Prats, K6ACJ; AMSAT and TAPR; John Johnston, W3BE; California
Rescue ARES Net members; Question Pool Committee members; WA6TWF Super
System members; Don Arnold, W6GPS; and Chip Margelli, K7JA.*

Copyright © 1988, 1991, 1994, 1998, 2000, 2004, 2007   Visit Master Publishing
Master Publishing, Inc.   on the Internet at:
6125 W. Howard Street   www.masterpublishing.com
Niles, Illinois 60714
847-763-0916
fax: 847-763-0918
e-mail: MasterPubl@aol.com
All Rights Reserved

REGARDING THESE BOOK MATERIALS
Reproduction, publication, or duplication of this book, or any part thereof, in any
manner, mechanically, electronically, or photographically, is prohibited without the
express written permission of the Publisher. For permission and other rights under this
copyright, write Master Publishing, Inc.

**The Author, Publisher and Seller assume no liability with respect to the use of the
information contained herein.**

Seventh Edition
    9   8   7   6   5   4   3   2   1

Table of Contents

QUESTION POOL NOMENCLATURE

The latest nomenclature changes and question pool numbering system recommended by the Volunteer Examiner Coordinator's Question Pool Committee (QPC) have been incorporated in this book. The General Class (Element 3) question pool has been rewritten at the high-school reading level. This question pool is valid from July 1, 2007 through June 30, 2011.

FCC RULES, REGULATIONS AND POLICIES

The NCVEC QPC releases revised question pools on a regular cycle, and issues deletions as necessary. The FCC releases changes to FCC rules, regulations and policies as they are implemented. This book includes the most recent information released by the FCC at the time this copy was printed.

Welcome to *General Class*!

The General Class license gives you privileges on all of the worldwide amateur radio bands for long-range skywave communications. These new bands will add to your Technician Class VHF and UHF privileges, too. With General Class, you will receive operating privileges on every ham radio band there is for voice, data, video, and CW.

As a General Class operator, you'll be able to travel the world and always stay in touch using skywave communications. Whether you sail the international waters of the South Seas, or cruise the highways of North America or Europe, you can stay in touch with your new General Class worldwide band privileges. More good news – recent international agreements allow you to operate in many foreign ports and countries without having to officially sign-up for a reciprocal license. We discuss these new agreements with Europe and South America inside this book.

As a General Class operator, you'll be able to help our hobby grow! I encourage you to join an all-General Volunteer Examination team to conduct Element 2 Technician Class written exams, and to actively recruit new ham radio operators to our hobby and service.

This new Seventh Edition of my book covers everything you need to know to pass the General Class Element 3 written examination. There no longer is a Morse code test required for any class of ham license! The Federal Communications Commission eliminated the Morse code test requirement effective February 23, 2007.

The question pool in this book is effective July 1, 2007 through June 30, 2011. The only prerequisite for your new General Class license is a valid Technician Class license or a recently-earned CSCE for Technician Element 2. And with the elimination of the Morse code 5 wpm test, this new *General Class* book will make your test preparation a breeze. We've completely reorganized the entire, new Element 3 General Class question pool to improve your learning experience and cut down on the study time required to pass the theory exam.

I am on the air daily, and I hope to talk to you soon on the worldwide General Class bands!

73

Gordon West, WB6NOA

About This Book

My book provides you with all of the study material you need to prepare yourself to pass the Element 3 General Class examination. The precise test questions and 4 distracters are in this book. We have reorganized all of the questions and answers into common subject groups for more logical learning. This will cut down on your study time and improve your knowledge of how to operate as a General Class ham!

The Federal Communications Commission completely eliminated Morse code testing as a requirement for an amateur radio license effective February 23, 2007. I always encourage our students to know the code, so this book has a complete chapter on fun ways to begin learning Morse code once you are on the air as a new General Class operator. Chapter 5 gives you sound advice on how to learn the Morse code. Even though there is no more code test, knowing the dots and dashes will help you become a better General Class operator.

My book also provides you with valuable information you need to be an active ham on the HF worldwide bands. To help you get the most out of *General Class*, here's a look at how the book is organized:

- *Chapter 1* explains in detail the new HF operating privileges you'll earn by passing your Element 3 General Class written theory examination.

- *Chapter 2* provides a look at the history of amateur radio licensing in the U.S., and the current status of all license classes.

- *Chapter 3* contains all 484 Element 3 General Class questions, 35 of which will be on your upcoming exam. You'll notice that the questions and answers have been reorganized to follow the syllabus I use to teach General Class at my Radio School weekend sessions. It makes your learning experience easier and more meaningful.

- *Chapter 4* tells you how to find an exam site, what to do before you get to the site, what required papers to bring with you, and what to check for when you pass your exam.

- *Chapter 5* gives sound advice on how to learn Morse code at 5 wpm.

- The *Appendix* contains a list of VECs, a Glossary of important ham radio terms, and other useful information.

If you need additional study materials, I've recorded an audio course that follows this book, and W5YI also has interactive software to allow you to study for upcoming exams on your PC. In addition, I also have recorded audio courses for learning Morse code. W5YI also offers Morse code software so you can use your PC to learn CW at 5-wpm, and higher.

Need additional study materials?
Call W5YI at 800-669-9594 or go to www.w5yi.com

1

General Class Privileges

The General Class amateur operator/primary station license permits you to use segments of every worldwide band for long-distance voice, data, and video amateur communications. You also keep all of the privileges you now enjoy as a Technician, or grandfathered Novice or Technician-Plus operator.

The General Class license has always been considered "the big one" because of the almost unlimited skywave privileges this license provides on each of the worldwide ham bands. Although the older, grandfathered Novice and Technician-Plus licenses offer slivers of worldwide band privileges, it is the General Class license that gives you major operating "elbow room" on segments of all the worldwide bands. Whether it's day or night, summer or winter, these worldwide bands will always offer skywave communications for thousands of miles of range.

Never in the history of amateur radio has it been easier than now to obtain the General Class license. There is no more Morse code test. The Federal Communications Commission concluded "...this change (eliminating the code test) eliminates an unnecessary burden that may discourage current amateur radio operators from advancing their skills and participating more fully in the benefits of amateur radio." Many other countries have already dropped their Morse code test requirements, too.

Gordo at the controls of the United Nations Station, 4U1UN, in New York.

The Element 3 theory exam has been dramatically improved in content, better reflecting General Class worldwide radio equipment and antennas. Simple math questions deal with modern, high frequency equipment. No longer will you need to memorize "far out" formulas that you would rarely use as a licensed ham radio operator. There is more emphasis on General Class operating, band plans, data modes, and high frequency simple antennas to build. Although this new question pool is more comprehensive in entire content, your upcoming exam remains at 35 multiple choice questions, 74% passing grade.

WORLDWIDE SPECTRUM

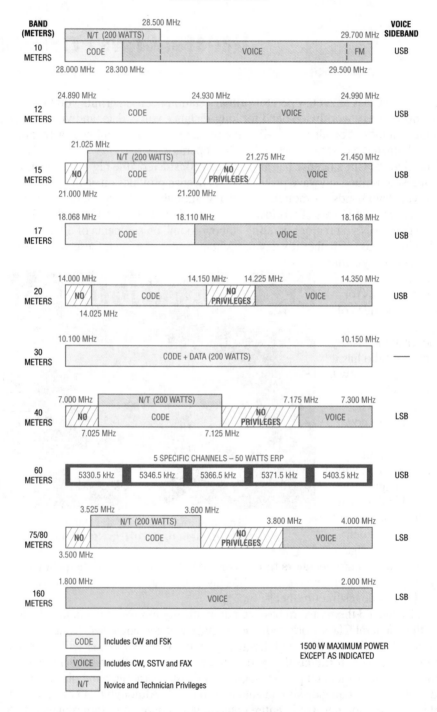

Figure 1-1. General Class HF License Privileges

GENERAL CLASS LICENSE PRIVILEGES

Figure 1-1 graphically illustrates your new General Class code, data, and voice privileges on the medium frequency (MF) bands (300 kHz-3 MHz), and high frequency (HF) bands (3 MHz-30 MHz). Code and data privileges are in the designated areas on the left side of each band. Voice privileges are on the right, upper side of each band. Designated areas between the code and voice privileges have "no privileges" for the General Class operator. These are reserved for grandfathered Advanced Class and current Extra Class operators.

As you can see, grandfathered Advanced and Extra Class operators have the same band privileges that you will enjoy as a General Class operator, they just have a little bit more elbow room. But don't worry – there is plenty of room throughout the General Class voice spectrum for working the world!

160 METERS, 1.8 MHZ - 2.0 MHZ

Your General Class privileges are the same as Advanced and Extra on this band. You may operate voice and code from one end to the other. The 160-meter band is great for long-distance, nighttime communications. It's located just above the AM broadcast radio frequencies. At night, 160 meters lets you work the world.

75/80 METERS, 3.5 MHZ - 4.0 MHZ

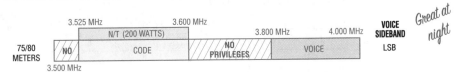

On 75/80 meters, General Class code, data, radio teleprinter (RTTY) privileges are from 3.525 to 3.6 MHz, and single sideband voice privileges were recently expanded and go from 3.8 to 4.0 MHz.

60 METERS – 5 CHANNELS

Our new 60-meter band was allocated by the FCC on July 3, 2003. Unlike the other ham bands where we have a frequency spectrum, our 60-meter privileges are assigned *5 specific channels* for upper sideband communications. Effective radiated power is limited to 50 watts, and the 5 USB channels buzz with activity. The dialed USB frequencies assigned to the amateur service are: 5330.5 kHz; 5346.5 kHz; 5366.5 kHz; 5371.5 kHz; and 5403.5 kHz.

40 METERS, 7.0 MHZ - 7.3 MHZ

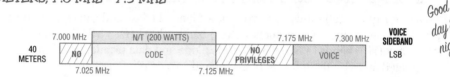

Good day & night

On 40 meters, General Class code, data, and RTTY privileges are from 7.025 to 7.125 MHz, and single sideband voice privileges recently were expanded and now go from 7.175 to 7.3 MHz. During daylight hours, 40 meters is a great band for base station and mobile contacts up to 500 miles away. At night, 40 meters skips all over the country, and many times all over the world. However, at nighttime, 40 meter worldwide foreign broadcast AM shortwave stations come booming in, giving you practice dodging all of the megawatt carriers.

30 METERS, 10.1 MHZ - 10.15 MHZ

Hot band for CW & data

Only code, data, and RTTY are permitted on this band. Thirty meters is located just above the 10-MHz WWV time broadcasts on shortwave radio. Voice is not allowed on this band by any class of amateur operator, and power is limited to 200 watts PEP.

20 METERS, 14.0 MHZ - 14.350 MHZ

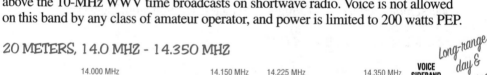

Long-range day & evening conta[...]

This is the best DX worldwide band there is, day or night! Morse code, data, and radioteleprinter (RTTY) privileges extend from 14.025 to 14.150 MHz. General Class voice privileges extend from 14.225 to 14.350 MHz. This is where the real DX activity takes place. Almost 24 hours a day, you should be able to work stations in excess of 5000 miles away with a modest antenna setup on the 20-meter band. If you are a mariner, most of the long range maritime mobile bands are within your privileges as a General Class operator. If you are into recreational vehicles (RV's), there are nets all over the country especially for you. The band "where it's at" is 20 meters when you want to work the world from your car, boat, RV, or home shack!

17 METERS, 18.068 MHZ - 18.168 MHZ

Best during the da[...]

17 METERS	18.068 MHz	18.110 MHz	18.168 MHz	VOICE SIDEBAND
	CODE	VOICE		USB

There is plenty of elbow room here with lots of foreign DX coming in day and night. Most new base antennas have 17 meters included. All emission types are authorized on this newer Amateur Radio band.

15 METERS, 21.0 MHZ - 21.450 MHZ

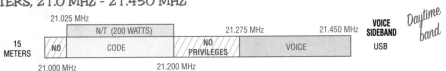

15 meter General Class CW, data and RTTY privileges extend from 21.025 to 21.200 MHz. Your single sideband voice privileges recently were expanded on 15 meters and go from 21.275 to 21.450 MHz. 15 meters is a great band for daytime skywave contacts, and 15 meters is a popular band for mobile operators because antenna requirements are relatively small. I have worked all over the world on the 15 meter band, mobile.

12 METERS, 24.890 MHZ - 24.990 MHZ

Code, data, and RTTY privileges on the 12-meter band extend from 24.890 MHz to 24.930 MHz. Voice privileges are from 24.930 to 24.990 MHz. You have the same privileges and elbow room as the Advanced and Extra Class operator, too. Although this is a very narrow band, expect excellent daytime range throughout the world. At night, range is limited to groundwave coverage because the ionosphere is not receiving sunlight to produce skywave coverage.

10 METERS, 28.0 MHZ - 29.7 MHZ

General Class CW, data, and RTTY privileges begin at the very bottom of the band, 28.0 MHz, and extend up to 28.3 MHz. Your voice privileges begin immediately at 28.300 MHz and extend up to 29.700 MHz. You have the same privileges on 10 meters as Advanced and Extra Class operators. Recent rule-making now allows all Technician Class operators voice privileges on 10 meters, but only from 28.3 to 28.5 MHz. This will be an exciting area to tune in when the band regularly opens for skywave contacts. This makes the 10 meter band a great spot to talk via skywaves with a new Technician Class operator, with plenty of room above 28.5 for exclusive General, Advanced and Extra Class privileges. And wait until you try the full fidelity of frequency modulation (FM) by giving a quick CQ call on 29.600 MHz, FM! There are even FM repeaters at the top of 10 meters, too.

Don't forget your handheld! When you upgrade to general, you keep your VHF/UHF/SHF privileges.

6 METERS AND UP

Your General Class license allows you unlimited band privileges and unlimited emission privileges on several higher-frequency bands, as shown in *Table 1-1*.

Table 1-1. 6 Meter and Higher Band Privileges

Frequency	Meters
50-54 MHz	6 meters
144-148 MHz	2 meters
222-225 MHz	1.25 meters
420-450 MHz	0.70 meters (70 cm)
902-928 MHz	0.35 meters (35 cm)
1240-1300 MHz	0.23 meters (23 cm)

MICROWAVE BANDS

Your General Class license allows you unlimited band privileges and unlimited emission privileges in the microwave bands, as indicated in *Table 1-2*. These VHF, UHF, and SHF frequencies are the same ones for which you received privileges when you passed your Technician or Technician-Plus Class examinations.

Table 1-2. Microwave Band Frequency Privileges

Frequency	Frequency
2300-2310 MHz	47.0-47.2 GHz
2390-2450 MHz	75.5-1.0 GHz
3.3-3.5 GHz	119.98-20.02 GHz
5.65-5.925 GHz	142-49 GHz
10.0-10.5 GHz	241-50 GHz
24.0-24.25 GHz	All above 300 GHz

My first book, *Technician Class*, for Element 2, provides a detailed explanation of the VHF, UHF, and SHF bands and presents the specific ARRL-recommended band plans for these frequencies. *Technician Class* is available from your local amateur radio dealer, at hamfests, or by calling The W5YI Group at 800/669-9594, or visit **www.w5yi.org**.

THE CONSIDERATE OPERATOR'S FREQUENCY GUIDE

Nothing in the FCC rules recognizes special privileges on any specific frequency for a net, group, or individual. No one "owns" a frequency. Rather, amateur operators rely on "gentlemen's agreements," good amateur practice, and common sense for all ham operators to check to see if the frequency is in use prior to transmitting – regardless of mode.

That said, here's a handy listing of frequencies that are generally recognized for certain modes or activities. My thanks to our friends at *QST* magazine, who gave us permission to reprint it from their March 2007 edition. Copyright © 2007, ARRL.

Frequency (MHz)	Mode / Activity
1.800 – 2.000	CW
1.800 – 1.810	Digital
1.810	QRP CW calling frequency
1.843-2.000	SSB, SSTV and other wideband modes
1.910	SSB QRP
1.995 – 2.000	Experimental
1.999 – 2.000	Beacons
3.500 – 3.510	CW DX window
3.560	QRP CW calling frequency
3.570 – 3.600	RTTY/Data
3.585 – 3.600	Automatically controlled data stations
3.590	RTTY/Data DX
3.790 – 3.800	DX Window
3.845	SSTV
3.885	AM calling frequency
3.985	QRP SSB calling frequency
7.030	QRP CW calling frequency
7.040	RTTY/Data DX
7.080 – 7.125	RTTY/Data
7.100 – 7.105	Automatically controlled data stations
7.070	PSK
7.285	QRP SSB calling frequency
7.290	AM calling frequency
10.130 – 10.140	RTTY/Data
10.140 – 10.150	Automatically controlled data stations
14.060	QRP SSB calling frequency
14.070 – 14.095	RTTY/Data
14.095 – 14.0995	Automatically controlled data stations
14.100	IBP/NCDXF beacons
14.1005 – 14.112	Automatically controlled data stations
14.230	SSTV
14.285	QRP SSB calling frequency
14.286	AM calling frequency

Frequency (MHz)	Mode / Activity
18.100 – 18.105	RTTY/Data
18.105 – 18.110	Automatically controlled data stations
18.110	IBP/NCDXF beacons
21.060	QRP CW calling frequency
21.070 – 21.110	RTTY/Data
21.090 – 21.100	Automatically controlled data stations
21.150	IBP/NCDXF beacons
21.340	SSTV
21.385	QRP SSB calling frequency
24.920 – 24.925	RTTY/Data
24.925 – 24.930	Automatically controlled data stations
24.930	IBP/NCDXF beacons
28.060	QRP CW calling frequency
28.070 – 28.120	RTTY/Data
28.120 – 28.189	Automatically controlled data stations
28.190 – 28.225	Beacons
28.200	IBP/NCDXF beacons
28.385	QRP SSB calling frequency
28.680	SSTV
29.000 – 29.200	AM
29.300 – 29.510	Satelite downlinks
29.520 – 29.580	Repeater inputs
29.600	FM simplex
29.620 – 29.680	Repeater outputs

TECHNICIAN EXAM ADMINISTRATION

There is one more very important privilege you gain when you achieve General Class status. As a General Class licensee, you may take part in the administration of Element 2 Technician Class written examinations, once you become accredited as a volunteer examiner at the General Class level by a VEC.

So, if you wish to see the amateur service grow in your local area, find 2 other General Class operators, then contact your local or national VEC for accreditation, and start your own testing team for newcomers to our hobby.

> To become an accredited Volunteer Examiner as a General Class licensee call the W5YI VEC at 800-669-9594.

SUMMARY

In late 2006, the FCC "refarmed" code, data, and voice privileges for General, Advanced, and Extra Class operators. All of the band charts in this new book have been updated to reflect the added privileges for high frequency voice operation. The FCC tightened up the code spectrum, and added more elbow room to the voice spectrum. General Class operators gained:

- 50 kHz of voice spectrum on 75 meters
- 50 kHz of voice spectrum on 40 meters
- 25 kHz of voice spectrum on 15 meters

With the elimination of the Morse code test on February 23, 2007, and the addition of all of this voice spectrum, General Class is the place to be!

Are you ready to prepare for the General Class Element 3 written examination? I sure hope so. Welcome – in advance – to the worldwide bands! I hope to hear you on the high frequency bands very soon.

2

A Little Ham History!

Ham radio has changed a lot in the 100-plus years since radio's inception. In the past 25 years, we have seen some monumental changes! So, before we get started preparing for the exam, I'm going to give you a little refresher lesson about our hobby, its history, and an overview of how you'll progress through the amateur ranks to the top amateur ticket – the Extra Class license. We know this background knowledge will make you a better ham! I'll keep it light and fun, so breeze through these pages.

In this chapter you'll learn all of the licensing requirements under the FCC rules that became effective April 15, 2000. And you'll learn about the six classes of license that were in effect *prior* to those rules changes. That way, when you run into a Novice, Technician Plus or Advanced Class operator on the air, you'll have some understanding of their skill level, experience, and frequency privileges.

WHAT IS THE AMATEUR SERVICE?

There are more than 650,000 Americans who are licensed amateur radio operators in the U.S. today. The Federal Communications Commission, the Federal agency responsible for licensing amateur operators, defines our radio service this way:

"The amateur service is for qualified persons of all ages who are interested in radio technique solely with a personal aim and without pecuniary interest."

Ham radio is first and foremost a fun hobby! In addition, it is a service. And note the word "qualified" in the FCC's definition – that's the reason why you're studying for an exam; so you can pass the exam, prove you are qualified, and get on the air.

Millions of operators around the world exchange ham radio greetings and messages by voice, teleprinting, telegraphy, facsimile, and television worldwide. Japan, alone, has more than a million hams! It is very commonplace for U.S. amateurs to communicate with Russian amateurs, while China is just getting started with its amateur service. Being a ham operator is a great way to promote international good will.

The benefits of ham radio are countless! Ham operators are probably known best for their contributions during times of disaster. In recent years, many recreational sailors in the Caribbean who have been attacked by modern-day pirates have had their lives saved by hams directing rescue efforts. Following the 9/11 terrorist attacks on the World Trade Center and the Pentagon, literally thousands of local hams assisted with emergency communications. After Katrina blew through New Orleans, ham radio operators were the first to report the levee breach and to warn that flood waters were rushing into the city.

The ham community knows no geographic, political or social barrier. If you study hard and make the effort, you are going to earn your upgrade to General Class and get on the HF worldwide bands. Follow the suggestions in my book and your chances of passing the written exam are excellent, and learning will be easy and fun!

A BRIEF HISTORY OF AMATEUR RADIO LICENSING

Before government licensing of radio stations and amateur operators was instituted in 1912, hams could operate on any wavelength they chose and could even select their own call letters. The Radio Act of 1912 mandated the first Federal licensing of all radio stations and assigned amateurs to the short wavelengths of less than 200 meters. These "new" requirements didn't deter them, however, and within a few years there were thousands of licensed ham operators in the United States.

Since electromagnetic signals do not respect national boundaries, radio is international in scope. National governments enact and enforce radio laws within a framework of international agreements that are overseen by the International Telecommunications Union. The ITU is a worldwide United Nations agency headquartered in Geneva, Switzerland. The ITU divides the radio spectrum into a number of frequency bands, with each band reserved for a particular use. Amateur radio is fortunate to have many bands allocated to it all across the radio spectrum.

In the U.S., the Federal Communications Commission is the government agency responsible for the regulation of wire and radio communications. The FCC further allocates frequency bands to the various services in accordance with the ITU plan – including the Amateur Service – and regulates stations and operators.

By international agreement, in 1927 the alphabet was apportioned among various nations for basic call sign use. The prefix letters K, N and W were assigned to the United States, which also shares the letter A with some other countries.

In the early years of amateur radio licensing in the U.S., the classes of licenses were designated by the letters "A," "B," and "C." The highest license class with the most privileges was "A." In 1951, the FCC dropped the letter designations and gave the license classes names. They also added a new Novice Class – a one-year, non-renewable license for beginners that required a 5-wpm Morse code speed proficiency test and a 20-question written examination on elementary theory and regulations, with both tests taken before one licensed ham.

In 1967, the Advanced Class was added to the Novice, Technician, General and Extra classes. The General exam required 13-wpm code speed, and Extra required 20-wpm. Each of the five written exams were progressively more comprehensive and formed what came to be known as the *Incentive Licensing System.*

In the '70s, the Technician Class license became very popular because of the number of repeater stations appearing on the air that extended the range of VHF and UHF mobile and handheld stations. It also was very fashionable to be able to patch your mobile radio into the telephone system, which allowed hams to make telephone calls from their automobiles long before the advent of cell phones.

In 1979, the international Amateur Service regulations were changed to permit all countries to waive the manual Morse code proficiency requirement for "...stations making use exclusively of frequencies above 30 MHz." This set the stage for the creation of the Technician "no-code" license, which occurred in 1991, when the 5-wpm Morse code requirement for the Technician Class was eliminated. New licensees were now permitted to operate on all amateur bands above 30 MHz. Applicants for the no-code Technician license had to pass the 35-question Novice and 30-question Technician Class written examinations but, for the first time, not a Morse code test. Technician Class amateurs who also passed a 5-wpm code test were awarded a Technician-Plus license. Besides their 30 MHz and higher no-code frequency privileges, Tech-Plus licensees gained the Novice CW privileges and a sliver of the 10 meter voice spectrum.

By this time, there was a total of six Amateur Service license classes – Novice, Technician, Technician-Plus, General, Advanced, and Extra – along with five written exams and three Morse code tests used to qualify hams for their various licenses.

The Amateur Service Is Restructured

As you can see, through the years the ham radio service underwent a myriad of changes. Rules were amended to meet new technology, and the FCC regularly received petitions for "small" rule making. Then, in 1998, the FCC didn't just come up with a new plan for ham radio – they begin working with the ham radio community to restructure our amateur radio service to make it more effective. The objective of this restructuring was to streamline the license process, eliminate unnecessary and duplicated rules, and reduce the emphasis on the Morse code test. The result of this review was a complete restructuring of the US amateur service, which became effective April 15, 2000. The Morse code test for General Class and Extra Class was dropped from 13- and 20-wpm respectively, down to-5 wpm for all. Since 2000, there have been only 3 amateur radio license classes:

- Technician Class, the VHF/UHF entry level license
- General Class, the HF entry level license, which required a 5-wpm code test
- Amateur Extra Class, a technically oriented senior license, based on-5wpm code credit

Individuals with licenses issued before April 15, 2000, have been grandfathered under the new rules. This means that Novice, Technician Plus, and Advanced Class amateurs are able to modify and renew their licenses indefinitely. Technician Plus amateur licenses will be renewed as "Technician Class," but their original 5 wpm credit allowed them HF privileges indefinitely.

The Granddaddy of all restructuring occurred on February 23, 2007, when the FCC completely eliminated any Morse code examination for all amateur radio licenses. The FCC action was based on 6,200 written comments, with most supporting the elimination of all code tests.

The FCC also "upgraded" No-Code Technician Class operators for Technician high frequency privileges, mostly code only, on 80 meters, 40 meters, 15 meters, and code, data and 200 kHz of voice spectrum on 10 meters.

In its public notice announcing elimination of the Morse code test requirement, the FCC Commissioners wrote: "We believe that because the international requirement for telegraphy proficiency has been eliminated, we should treat Morse code telegraphy no differently than other amateur service communication technique. This change eliminates an unnecessary regulatory burden that may discourage current amateur radio operators from advancing their skills and participating more fully in the benefits of amateur radio."

Eliminating the Morse code test in no way diminishes the enthusiasm many ham operators have for the technique of sending dits and dahs over the air. In fact, I believe we will have more General Class hams learning the code than ever before, now that they can do it on high frequency where practicing code with other hams is fun, even though they begin by using only the more simple code characters along with their call signs!

Self-Testing In The Amateur Service

Prior to 1984, all amateur radio exams were administered by FCC personnel at FCC Field Offices around the country. In 1984, the FCC adopted a two-tier system beneath it called the VEC System to handle amateur radio license exams. It also increased the length of the term of amateur radio licenses from five to ten years.

The VEC (Volunteer Examiner Coordinator) System was formed after Congress passed laws that allowed the FCC to accept the services of Volunteer Examiners (or VEs) to prepare and administer amateur service license examinations. The testing activity of VEs is managed by Volunteer Examiner Coordinators (or VECs). A VEC acts as the administrative liaison between the VEs who administer the various ham examinations and the FCC, which grants the license.

A team of three VEs, who must be approved by a VEC, is required to conduct amateur radio examinations. General Class amateurs may serve as examiners for the Technician class. Advanced Class amateurs may administer exams for Elements 2 and 3. The Extra Class written Element 4 may only be administered by a VE who holds an Extra Class license.

In 1986, the FCC turned over responsibility for maintenance of the exam questions to the National Conference of VECs, which appoints a Question Pool Committee (QPC) to develop and revise the various question pools according to a schedule. The QPC is required by the FCC to have at least ten times as many questions in each of the pools as may appear on an examination. As a rule, each of the 3 question pools is changed once every 4 years.

The previously-mandated ten written exam topic areas were eliminated and the Question Pool Committee now decides the content of each of the three written examinations. Both the Technician Class Element 2 and General Class Element 3 written examinations contain 35 multiple-choice questions. The Extra Class Element 4 written examination has 50 questions.

OPERATOR LICENSE REQUIREMENTS

To qualify for an amateur operator/primary station license, a person must pass an examination according to FCC guidelines. The degree of skill and knowledge that the candidate demonstrates to the examiners determines the class of operator license for which the person is qualified.

Anyone is eligible to become a U.S. licensed amateur operator (including foreign nationals, if they are not a representative of a foreign government). There is no age limitation – if you can pass the examinations, you can become a ham!

OPERATOR LICENSE CLASSES AND EXAM REQUIREMENTS

Today, there are three amateur operator licenses issued by the FCC – Technician, General, and Extra. Each license requires progressively higher levels of learning and proficiency, and each gives you additional operating privileges. This is known as *incentive licensing* – a method of strengthening the amateur service by offering more radio spectrum privileges in exchange for more operating and electronic knowledge.

There is no waiting time required to upgrade from one amateur license class to another, nor any required waiting time to retake a failed exam. You can even take all three examinations at one sitting if you're really brave! *Table 2-1* details the amateur service license structure and required examinations.

Table 2-1: Current Amateur License Classes and Exam Requirements

License Class	Exam Element	Type of Examination
Technician Class	2	35-question, multiple-choice written examination. Minimum passing score is 26 questions answered correctly (74%).
General Class	3	35-question, multiple-choice written examination. Minimum passing score is 26 questions answered correctly (74%).
Extra Class	4	50-question, multiple-choice written examination. Minimum passing score is 37 questions answered correctly (74%).

ABOUT THE WRITTEN EXAMS

What is the focus of each of the written examinations, and how does it relate to gaining expanding amateur radio privileges as you move up the ladder toward your Extra Class license? *Table 2-2* summarizes the subjects covered in each written examination element.

Table 2-2. Question Element Subjects

Exam Element	License Class	Subjects
Element 2	Technician	Elementary operating procedures, radio regulations, and a smattering of beginning electronics. Emphasis is on VHF and UHF operating.
Element 3	General	HF (high-frequency) operating privileges, amateur practices, radio regulations, and a little more electronics. Emphasis is on HF bands.
Element 4	Extra	Basically a technical examination. Covers specialized operating procedures, more radio regulations, formulas and heavy math. Also covers the specifics on amateur testing procedures.

No Jumping Allowed

All written examinations for an amateur radio license are additive. You *cannot* skip over a license class or by-pass a required examination as you upgrade from Technician to General to Extra. For example, to obtain a General Class license, you must first take and pass the Element 2 written examination for the Technician Class license, plus the Element 3 written examination. To obtain the Extra Class license, you must first pass the Element 2 (Technician) and Element 3 (General) written examinations, and then successfully pass the Element 4 (Extra) written examination.

TAKING THE ELEMENT 3 EXAM

Here's a summary of what you can expect when you go to the session to take the Element 3 written examination for your General Class license. Detailed information about how to find an exam session, what to expect at the session, what to bring to the session, and more, is included in Chapter 4.

Examination Administration

All amateur radio examinations are administered by a team of at least three Volunteer Examiners (VEs) who have been accredited by a Volunteer Examiner Coordinator (VEC). The VEs are licensed hams who volunteer their time to help our hobby grow.

How to Find an Exam Session

Examination sessions are organized under the auspices of an approved VEC. A list of VECs is located in the Appendix on page 200. The W5YI-VEC and the ARRL-VEC are the 2 largest examination groups in the country, and they test in all 50 states. Their 3-member, accredited examination teams are just about *everywhere*. So when you call the VEC, you can be assured they probably have an examination team only a few miles from where you are reading this book right now!

Want to find a test site fast?
Visit the W5YI-VEC website at **www.w5yi.org**, or call 800-669-9594.

Taking the Exam

The Element 3 written exam is a multiple-choice format. The VEs will give you a test paper that contains the 35 questions and multiple choice answers, and an answer sheet for you to complete. Take your time! Make sure you read each question carefully and select the correct answer. Once you're finished, double check your work before handing in your test papers.

The VEs will score your test immediately, and you'll know before you leave the exam site whether you've passed. Chances are very good that, if you've studied hard, you'll get that passing grade!

GETTING/KEEPING YOUR CALL SIGN

Once the VE team scores your test and you've passed, the process of getting your official FCC Amateur Radio License begins – usually that same day.

At the exam site, you will complete NCVEC Form 605, which is your application to the FCC for your license. If you pass the exam, the VE team will send on the required paperwork to their VEC. The VEC reviews the paperwork submitted by your exam team and then files your application with the FCC. This filing is done electronically, and your license upgrade will be granted and posted on the FCC's website within a few days.

If you didn't check the "Change Call Sign" box on your NCVEC Form 605 application, you will simply keep your current call sign. However, if you do check the "Change Call Sign" box, you will receive an entry-level "Group D" call sign as if you were a newly-licensed amateur. I suggest that you don't check the "Change Call Sign" box and stick with your present call sign.

As soon as you have your CSCE, you are permitted to go on the air with your General Class HF privileges – even before your paper license arrives in the mail! See Chapter 4 for more details on this process.

Vanity Call Signs

Your first call sign is assigned by the FCC's computer, and you have no choice of letters. However, once you have that call sign, you can apply for a Vanity Call Sign. Again, see Chapter 4 for details.

HOW MANY CLASSES OF LICENSES?

Once you've passed your Element 3 exam and go on the air as a new General Class operator, you'll be talking to fellow hams throughout the U.S. and around the world. Here's a summary of the new and "grandfathered" licenses that your fellow amateurs may hold, and a recap of the level of expertise they have demonstrated in order to gain their licenses.

New License Classes

Following the FCC's restructuring of Amateur Radio that took effect April 15, 2000, there are just three written exams and three license classes – Technician, General, and Extra. But persons who hold licenses issued prior to April 15, 2000, may continue to hold onto their license class and continue to renew it every 10 years for as long as they wish.

"Grandfathered" Licensees

As mentioned previously, individuals licensed prior to April 15, 2000, will continue to enjoy band privileges based on their licenses. So, once you get on the air with your new General Class privileges, every now and then you might meet a Novice operator while yakking on 10 meters, or sending CW on 15, 40, or 80 meters.

And you'll see some older licenses saying "Technician Plus," which belong to grandfathered Technician class operators licensed prior to April 15, 2000, who passed their 5-wpm code test and who get to keep their code credit indefinitely as long as they renew their license. Technician Plus operators will have their licenses renewed as Technician class.

And then there are the Advanced Class operators who may continue to hold onto their license class designation until they finally decide to move up to Extra Class.

When you look at the brand new Frequency Charts in this book, you'll see that we have updated them to reflect the expanded privileges on many high frequency bands for General Class operators and higher. We also list the sub-band privileges for Extra Class, Advanced Class, General Class, Technician Class and even the Novice. Look for my poster-sized, color-coded chart offered by many ham radio manufacturers that shows all of these operating privileges.

IT'S EASY!

Probably the primary pre-requisite for passing any amateur radio operator license exam is the will to do it. If you follow my suggestions in this book, your chances of passing the General exam are excellent. And once you pass, then it's on to our *Extra Class* book.

Yes, indeed, 2006 and 2007 FCC rulemaking has opened up all of the high frequency bands for expanded voice privileges, and General Class licensing exams without a Morse code test.

When you become a General Class operator, GET ON THE AIR! Sure, pick up my *Extra Class* book, but start operating as an important step to becoming a good ham and ultimately a candidate for the highest license – Amateur Extra Class. But, GET ON THE AIR as a General Class ham!

Gordo and Chip work 500 miles on 432 MHz using just 30 watts!

3

Getting Ready for the Exam

Your General Class written examination will consist of 35 multiple-choice questions taken from the 484 questions that make up the 2007-11 Element 3 pool. Each question on your examination and the multiple-choice answer will be identical to what is contained in this book.

This chapter contains the official, complete 484-question FCC Element 3 General Class question pool from which your examination will be taken. Again, your exam will contain 35 of these questions, and you must get 74% of the questions correct – which means you must answer 26 questions correctly in order to pass. This chapter also contains important information about how the exam is constructed using the questions taken from the Element 3 pool.

Your examination will be administered by a team of 3 or more Volunteer Examiners (VEs) – amateur radio operators who are accredited by a Volunteer Examiner Coordinator (VEC). You will receive a Certificate of Successful Completion of Examination (CSCE) when you pass the examination. This is official proof that you have passed the exam and it will be given to you before you leave the exam center.

THE 2007-11 QUESTION POOL

The Element 3 General Class question pool contained in this book is valid from July 1, 2007, through June 30, 2011.

The 484 questions and distracters in the new General Class question pool were developed by the National Conference of Volunteer Examiner Coordinators' Question Pool Committee (NCVEC-QPC). Committee members are Jim Wiley, KL7CC, chairman, Roland Anders, K3RA, Perry Green, WY1O, and Larry Pollock, NB5X.

The QPC encourages amateur radio operators throughout the country to submit revised questions for the amateur radio pools to the committee. If you have any suggestions for new or revised questions, you can send them to me, and I'll be happy to forward them on to the QPC. My address is on page 189.

WHAT THE EXAMINATION CONTAINS

The examination questions and the multiple-choice answers (one correct answer and three "distracters") for all license class levels are public information. They are widely published and are identical to those in this book. *There are no "secret" questions.* FCC rules prohibit any examiner or examination team from making any changes to any questions, including any numerical values. No numbers, words, letters, or punctuation marks can be altered from the published question pool. By studying all 484 Element 3 questions in this book, you will be reading the same exact questions that will appear on your 35-question Element 3 written examination. But which 35 out of the 484 total questions?

Subelement	Topic	Total Questions	Exam Questions
	Table 3-1. FCC Element 3 General Class Question Pool		
G1	Commission's Rules	65	5
G2	Operating Procedures	71	6
G3	Radio Wave Propagation	47	3
G4	Amateur Radio Practices	66	5
G5	Electrical Principles	44	3
G6	Circuit Components	41	3
G7	Practical Circuits	38	2
G8	Signals and Emissions	24	2
G9	Antennas and Feedlines	59	4
G0	Electrical and RF Safety	29	2
TOTALS		484	35

Table 3-1 shows how the Element 3, 35 question examination will be constructed. The question pool is divided into 10 sub-elements. Each sub-element covers a different subject, and is divided into topic areas. For example, for the Element 3 examination, two questions from the 35 total exam questions will be taken from sub-element G 8, Signals and Emissions. On sub-element G1, Commission's Rules, you will have 5 exam questions on your test. Rusty on electrical principles? Your test will have only 3 questions from sub-element G 5.

All Volunteer Examination teams use the same multiple-choice question pool. This uniformity in study material ensures common examinations throughout the country. Most exams are computer-generated, and the computer selects one question from each topic area within each subelement for your upcoming Element 3 exam.

Trust me – trust me, every question on your upcoming Element 3 exam will look very familiar to you by the time you finish studying this book

QUESTION CODING

Each and every question in the 484 question Element 3, General Class pool is numbered using a **code**. *The coded numbers and letters reveal important facts about each question!*

The numbering code always contains 5 alphanumeric characters to identify each question. Here's how to read the question number so you know exactly how the examination computer will select one question out of each group for your exam. Once you know this information, you can increase your odds of achieving a "max" score on the exam, especially if there is a specific group of questions which seems impossible for you to memorize or understand. When you get to Element 4, Extra Class – a very tough exam – this trick will really come in handy!

Let's pick a typical question out of the pool – G1A04 – and let me show you how this numbering code works:

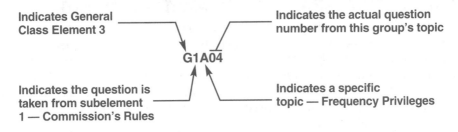

Figure 3-1. Examination Question Coding

- The first character "G" identifies the license class question pool from which the question is taken. "T" would be for Technician. "G" is for General, and "E" would be for Extra.
- The second digit, a "1", identifies the subelement number, 1 through 0. General subelement 1 deals with FCC Rules.
- The third character, "A", indicates the topic area within the subelement. Topic "A" deals with your General Class frequency privileges.
- The fourth and fifth digits indicate the actual question number within the subelement topic's group. The "04" indicates this is the fourth question about frequency privileges, and within the topic area of Commission's Rules. There are 16 individual questions in topic area G1A, *but only one question out of this topic group will appear on the test.*

Here's the Secret Study Hint

Only one exam question will be taken from any single group! A computer-generated test is set up to take one question from one single topic group. It cannot skip any one group, nor can it take any more than one question from that group.

Your upcoming Element 3, General Class written exam is relatively easy with no bone-crusher math formulas. But when you get to the *Extra Class* book, there may be one or two groups that have formulas so complicated that you may want to wait until the very end to digest them. And if you decide to skip them completely, guess what – how many questions out of any one group? That's right, only one per group. This means you are not going to get hammered on any upcoming test with a whole bunch of questions dealing with a specific topic. Great secret, huh?

Study Time

How long will it take you to prepare for your upcoming exam?

The General Class question pool in this book is valid from July 1, 2007, through June 30, 2011. It contains a total of 484 questions – but don't panic! Most questions are repeated several different ways, and these "repeats" reinforce what you've already learned. It is probably going to take about 30 days to work through this book and prepare for your General Class exam. Remember, you must take the written exams in order: Element 2 Technician, Element 3 General, and Element 4 Extra.

QUESTIONS REARRANGED FOR SMARTER LEARNING

The first thing you'll notice when you look at how the Element 3 question pool is presented in this book is that I have *completely rearranged the entire General Class question pool* to precisely follow my weekend ham radio training classes. This rearrangement will take you *logically* through each and every one of the 484 Q & A's. The questions are arranged here by 18 topic areas in a way that eliminates the need for you to jump back and forth between topic groups or subelements to match up questions on similar topics.

For example, I have taken all of the questions in the pool that talk about where you can operate your ham radio and grouped them together into one area that allows you to better understand all of the material that relates to this topic. This arrangement of the Q&A's follows a natural learning process beginning with your new General Class privileges, participating on a VE team, and getting on the air, and ending with questions on antennas, feed lines, and RF saefty.

Trust me, the reorganization of all of the test questions in the 484 pool has been tested and finely-tuned in hundreds of my weekend classes. This method of learning WORKS! You will probably cut the amount of study time in half simply by following the question pool from front to back as presented here in my book!

Let me assure that each and every Element 3 Q&A is in this book. A cross-reference of all 484 questions is found on pages 211 to 212 in the back of the book, along with the syllabus used by the Question Pool Committee to develop this new Element 3 pool.

This book – and my General Class audio course – contain all 484 General Class questions, 4 possible answers, the noted correct answer, and my upbeat description of how the correct answer works into the real world of amateur radio. We highlight **KEY WORDS** that will help you remember the correct answer and provide you with a fast review of the entire question pool just before you sit for the big General exam.

We also include many web addresses that can provide you with hours of fascinating study on selected "hot topics" that will help you really understand the real world of ham radio behind the Q & A's. You can visit the site while you study, or visit them after you've earned your General Class license and are on the air.

When you visit some of these websites, it may not be immediately apparent why we are suggesting that you go there. Some addresses take you to the sites of local ham clubs. Most of these are specialty clubs, and they contain lots of information on how to operate on repeaters, or satellites, or provide educational resources on learning about electronics or antennas or how radios work.

And a disclaimer – while we have worked hard to make sure all of these addresses are current at the time of publication, websites move, addresses change, or sites simply go away. So if you find an address that doesn't work, feel free to drop us an e-mail so we can update this for the next printing of our book. And if you know of a website that you think is a gem, then send us that information and we'll consider it for our next printing.

Here's our e-mail address:
masterpubl@aol.com

How to Read the Questions

Using an actual question from page 23, here is a guide to explain what it is you will be studying as you go through all of the Q&As in the book:

G1D08 What document must be issued to a person that passes an exam element?

Official Q&A

A. FCC form 605   C. CCSA

B. CSCE   D. NCVEC form 605

Key Words To Remember — Volunteer examiners will issue a *Certificate of Successful Completion of Examination (CSCE)*, signed by 3 VEs, indicating you passed your General Class test with flying colors! [97.509(i)] **ANSWER B**

FCC Part 97 Rule Citation   Correct Answer

Topic Areas

Here is a list of my 18 topic areas showing the page where it starts in the book. Again, there is a complete cross reference list of the Q&As in numerical order on pages 211 to 212 in the Appendix, along with the official Question Pool Syllabus.

Your Passing CSCE	23	Your HF Transmitter	86
Your New General Bands	25	Your Receiver	100
FCC Rules	34	Oscillators & Components	106
Be a VE!	41	Electrical Principles	117
Voice Operation	45	Circuits	129
CW Lives!	52	Good Grounds	140
Digital Operating	57	HF Antennas	145
In an Emergency	65	Coax Cable	165
Skywave Excitement!	71	RF & Electrical Safety	172

STUDY SUGGESTIONS

Finally, as you get ready to start studying the questions, here are some suggestions to make your learning easier:

1. Read over each multiple-choice answer carefully. Some answers start out looking good, but turn bad during the last 2 or 3 words. If you speed read the answers, you could very easily go for a wrong answer because you didn't read them all the way through. Also, don't count on the multiple-choice answers always appearing in the exact same A-B-C-D order on your actual computer-generated test. While they won't change any words in the answers, they will sometimes scramble the A-B-C-D order.

2. Keep in mind that there is only one question on your test that will come from each group, and track how many groups in each sub-element.

3. Give this book to a friend, and ask him or her to read you the correct answer. You then reply with the question wording.

4. Mark the heck out of your book! When the pages begin to fall out, you're probably ready for the exam!

5. Take this book with you everywhere you go. Put a check mark on those questions you have mastered. Use my four CDs audio course to learn some shortcuts to problem solving. The CD audio course is a great way to study while driving!

6. Highlight the keywords one week before the test. Then speed read the brightly highlighted key words twice a day before the exam.

Are you ready to work through the 484 Q & A's? Put a check mark by the easy ones that you may already know the answer for, and put a little circle by any question that needs a little bit more study. Save your highlighting work until a few days before your upcoming test. Work the Q & A's for about 30 minutes at a time. I'll drop in a little bit of humor to keep you on track; and if you actually need my live words of encouragement, you can call me on the phone Monday through Thursday, 10:00 a.m. to 4:00 p.m. (California time), 714-549-5000.

THE QUESTION POOL, PLEASE

Okay, this is the big moment – your General Class, Element 3, question pool. Don't freak out and get overwhelmed with the prospect of learning 484 Q & A's. You will find that the topic content is repeated many times, so you're really going to breeze through this test without any problem!

G1D08 What document must be issued to a person that passes an exam element?

A. FCC form 605   C. CCSA

B. CSCE   D. NCVEC form 605

Volunteer examiners will issue a *Certificate of Successful Completion of Examination (CSCE)*, signed by 3 VEs, indicating you passed your General Class test with flying colors! [97.509(i)] **ANSWER B**

G1D09 How long is a Certificate of Successful Completion of Examination (CSCE) valid for exam element credit?

A. 30 days

B. 180 days

C. 365 days

D. For as long as your current license is valid

You pass your test, you are holding your *CSCE*, and four months later you learn your passing paperwork was somehow lost in the mail. No problem – your CSCE is a valid exam element credit document, and is good for *365 days*. [97.3(a)(15)] **ANSWER C**

Don't Lose Your CSCE – It Is Proof of Your Privileges Until Your License Arrives

---

G1D06 When must you add the special identifier "AG" after your call sign if you are a Technician Class licensee and have a CSCE for General Class operator privileges?

A. Whenever you operate using General Class frequency privileges

B. Whenever you operate on any amateur frequency

C. Whenever you operate using Technician frequency privileges

D. A special identifier is not required as long as your General Class license application has been filed with the FCC

Until your call sign appears on the Universal Licensing System data base, *add the* special identifier *"temporary AG"* when you operate on your new General Class privileges. You don't need to append this identifier when operating on your current Technician frequency privileges – just your new General Class frequencies. [97.119(f)(2)] **ANSWER A**

G1D01 What is the proper way to identify when transmitting on General Class frequencies if you have a CSCE for the required elements but your upgrade from Technician has not appeared in the ULS database?

A. Give your call sign followed by the words "General Class"

B. No special identification is needed, since your license upgrade would already be shown in the FCC's database

C. Give your call sign followed by the words "temporary AG"

D. Give your call sign followed the abbreviation "CSCE"

Just earned your General Class CSCE? Go ahead and get on the air!

When you upgrade from Technician Class to General Class it will take about 10-15 days for your upgrade to appear on the Universal Licensing System database. But you don't need to wait to go on the air, because you have a completed CSCE (Certificate of Successful Completion of Examination) that allows you to *give your current Technician Class call sign*, followed by the words *"temporary AG."* Everyone will welcome you as a new General Class operator on the air when they hear "temporary AG." [97.119(f)(2)] **ANSWER C**

G1D03 Which of the following band segments may you operate on if you are a Technician Class operator and have a CSCE for General Class privileges?

A. Only the Technician band segments until your upgrade is posted on the FCC database

B. Only on the Technician band segments until your license arrives in the mail

C. On any General Class band segment

D. On any General Class Band segment except 30 and 60 meters

Once you pass your General Class written examination, you are good to go on the air *on all of the worldwide bands*, plus your present Technician frequencies, immediately with the CSCE. Keep my book handy to spot the operating segments for *General Class* operation on each band. [97.9(b)] **ANSWER C**

G1E09 What language must you use when identifying your station if you are using a language other than English in making a contact?

A. The language being used for the contact

B. Any language if the US has a third party agreement with that country

C. English

D. Any language of a country that is a member of the ITU

If you are allowing third party traffic to pass through your station from another country, double check that we have a third party agreement, and *identify* your own station only *using English.* [97.119(b)(2)] **ANSWER C**

 **Elmer Point:** *Here are the formulas that you need to convert frequency to wavelength, and wavelength to frequency:*

Converting Frequency to Wavelength

To find wavelength (λ) in meters, if you know frequency (f) in megahertz (MHz), Solve:

$$\lambda(meters) = \frac{300}{f(MHz)}$$

Converting Wavelength to Frequency

To find frequency (f) in megahertz (MHz), if you know wavelength (λ) in meters, Solve:

$$f(MHz) = \frac{300}{\lambda(meters)}$$

G1A01 On which of the following bands is a General Class license holder granted all amateur frequency privileges?

    A. 20, 17, and 12 meters
    B. 160, 80, 40, and 10 meters
    C. 160, 30, 17, 12, and 10 meters
    D. 160, 30, 17, 15, 12, and 10 meters

As a new General, you will gain High Frequency privileges on each and every ham band. But until you upgrade to Extra Class, you don't necessarily get ALL of that band's frequency privileges. On 80-, 40-, 20-, and 15-meters, Extra Class and Advanced Class operators have additional voice privileges. But not to worry – as a new General Class operator, you will receive ALL frequency privileges on *160, 30, 17, 15, 12, and 10 meters.* [97.301(d)] **ANSWER C**

G1A11 Which of the following frequencies is available to a control operator holding a General Class license?

A. 28.020 MHz

B. 28.350 MHz

C. 28.550 MHz

D. All of these answers are correct

As a General Class operator, you receive all privileges from 28.0 MHz to 29.7 MHz on the 10 meter band. *All* of the answers are *correct*. [97.301(d)] **ANSWER D**

G1C05 What is the maximum transmitting power a station with a General Class control operator may use on the 28 MHz band?

A. 100 watts PEP output

B. 1000 watts PEP output

C. 1500 watts PEP output

D. 2000 watts PEP output

Although Novice and Technician Class operators are restricted to 200 watts on High Frequency bands, as a General Class operator you could run up to *1500 watts* output. [97.313] **ANSWER C**

G1E02 When may a 10 meter repeater retransmit the 2 meter signal from a station having a Technician Class control operator?

A. Under no circumstances

B. Only if the station on 10 meters is operating under a Special Temporary Authorization allowing such retransmission

C. Only during an FCC-declared general state of communications emergency

D. Only if the 10 meter control operator holds at least a General Class license

10 meter repeaters may only operate between 29.5 and 29.7 MHz. Technician Class operators have no privileges on these frequencies. However, as a General, you can set up your home station as a cross-band relay system. This could allow a Technician on the 2-meter band to end up transmitting and receiving on the 10-meter band. Think of all the excitement you can give Technician Class operators on 2 meters when the 10-meter band is open for worldwide skywave communications. This is perfectly legal to do providing you stay at the control point of your station. You may never leave the relay station unattended without a *General Class*, or higher, *control operator* on active duty at the control point. [97.205(a)] **ANSWER D**

G1A06 Which of the following frequencies is in the 12 meter band?

A. 3.940 MHz

B. 12.940 MHz

C. 17.940 MHz

D. 24.940 MHz

Your new General Class 12 meter band extends from 24.890 MHz to 24.990 MHz, and there is only one answer with 24 MHz. Remember, to convert meters to megahertz, or megahertz to meters, 300 divided by either megahertz or meters will get you close! 300 divided by 12 = 25, which is close enough for the *24.940 MHz* answer.[97.301(d)] **ANSWER D**

G1C02 What is the maximum transmitting power an amateur station may use on the 12 meter band?
   A. 1500 PEP output, except for 200 watts PEP output in the novice portion
   B. 200 watts PEP output
   C. 1500 watts PEP output
   D. Effective radiated power equivalent to 50 watts from a half wave dipole

On the *12 meter band*, you can run the legal limit to *1500 watts PEP output*. [97.313(a)(b)] **ANSWER C**

G1A10 Which of the following frequencies is within the General Class portion of the 15 meter band?
   A. 14250 kHz
   B. 18155 kHz
   C. 21300 kHz
   D. 24900 kHz

The 15 meter band is at 21 MHz (21000 kHz), so *21300 kHz* is great place for you to operate voice to regularly work the world in mornings and afternoon. [97.301(d)] **ANSWER C**

G1A08 Which of the following frequencies is within the General Class portion of the 20 meter phone band?
   A. 14005 kHz
   B. 14105 kHz
   C. 14305 kHz
   D. 14405 kHz

Remember that phone privileges for High Frequency General Class operation are those at the top of the band. The *20 meter band* ends at *14350 kHz*, so 14305 kHz is within General Class voice privileges. [97.301(d)] **ANSWER C**

G1C04 What limitations, other than the 1500 watt PEP limit, are placed on transmitter power in the 14 MHz band?
   A. Only the minimum power necessary to carry out the desired communications should be used
   B. Power must be limited to 200 watts when transmitting between 14.100 MHz and 14.150 MHz
   C. Power should be limited as necessary to avoid interference to another radio service on the frequency
   D. Effective radiated power cannot exceed 3000 watts

Although you might be permitted to run an amplifier with 1500 watts output, *always try to run the minimum power* necessary to make contact with the other station. This will keep you out of your neighbors' TV and HiFi system. [97.313] **ANSWER A**

G1A02 On which of the following bands is phone operation prohibited?

A. 160 meters   C. 17 meters
B. 30 meters   D. 12 meters

As a new General, you can operate throughout the entire 30 meter band using CW and DATA emissions, 300 baud, with power output limited to 200 watts. *No voice* (phone) allowed *on 30 meters.* [97.305] **ANSWER B**

```
                10.100 MHz                                                        10.150 MHz   VOICE
                                                                                              SIDEBAND
    30        ┌──────────────────────────────────────────────────────────────┐
    METERS    │              CODE + DATA (200 WATTS)                          │      ──
              └──────────────────────────────────────────────────────────────┘
```

G1A03 On which of the following bands is image transmission prohibited?

A. 160 meters   C. 20 meters
B. 30 meters   D. 12 meters

Image transmissions include popular slow scan television and facsimile. On *30 meters, image transmissions are prohibited.* [97.305] **ANSWER B**

G1C01 What is the maximum transmitting power an amateur station may use on 10.140 MHz?

A. 200 watts PEP output   C. 1500 watts PEP output
B. 1000 watts PEP output   D. 2000 watts PEP output

On the *30 meter* CW and Data band, we can only run *200 watts* peak envelope power output to conform to FCC rules. [97.313(c)(1)] **ANSWER A**

G1A05 Which of the following frequencies is in the General Class portion of the 40 meter band?

A. 7.250 MHz   C. 40.200 MHz
B. 7.500 MHz   D. 40.500 MHz

Welcome to 40 meters where we recently gained an additional 50 kHz of voice spectrum, 7175 kHz to 7300 kHz lower sideband. Look for me most weekday mornings on *7.250 MHz*, within the General Class portion of the *40 meter band.* [97.301(d)] **ANSWER A**

```
            7.000 MHz                                  7.175 MHz   7.300 MHz   VOICE
                      ┌──── N/T (200 WATTS) ────┐                             SIDEBAND
    40      ┌──//──┬────────────────────┬──//────────┬────────────┐
    METERS  │//NO//│       CODE         │//  NO   // │   VOICE    │    LSB
            └──//──┴────────────────────┴//PRIVILEGES┴────────────┘
               7.025 MHz              7.125 MHz
```

G1C03 What is the maximum transmitting power a General Class licensee may use when operating between 7025 and 7125 kHz?

A. 200 watts PEP output   C. 1000 watts PEP output
B. 1500 watts PEP output   D. 2000 watts PEP output

General operators may operate CW and DATA from *7025-7125 kHz with 1500 watts* PEP output. [97.313] **ANSWER B**

G1A04 Which amateur band restricts communication to specific channels, using only USB voice, and prohibits all other modes, including CW and data?

A. 11 meters   C. 30 meters
B. 12 meters   D. 60 meters

We gained a new ham band, *60 meters*, shared with a few government stations on a non-interference basis – they can interfere with us, but we cannot interfere with them! We are authorized *5 discreet, upper sideband voice channels*, and these are great frequency assignments for passing regional radio traffic. 60 meters is the only band where we have been assigned specific channel allocations, and we may only use upper sideband voice on each of these channels. No Morse code, and no data, either. [97.303(s)] **ANSWER D**

	5 SPECIFIC CHANNELS – 50 WATTS ERP					VOICE SIDEBAND
60 METERS	5330.5 kHz	5346.5 kHz	5366.5 kHz	5371.5 kHz	5403.5 kHz	USB

G1C13 What is the maximum bandwidth permitted by FCC rules for amateur radio stations when operating on USB frequencies in the 60-meter band?

A. 2.8 kHz
B. 5.6 kHz
C. +/-2.8 kHz
D. 3 kHz

On the *60-meter* band, there are 5 channels allocated on specific frequencies where *maximum bandwidth is 2.8 kHz*. This is 2.8 kHz upper sideband, NOT upper (+) and lower (-). It is a good idea to keep your mike gain turned relatively low to prevent spilling out of this slightly-more-narrow bandwidth for SSB. Watch out for Answer C! [97.303(s)] **ANSWER A**

G1C07 Which of the following is a requirement when a station is transmitting on the 60 meter band?

A. Transmissions may only use Lower Sideband (LSB)
B. Transmissions must use only CW or Data modes
C. Transmissions must not exceed an effective radiated power of 50 Watts PEP referred to a dipole antenna
D. Transmissions must not exceed an effective radiated power of 200 Watts PEP referred to a dipole antenna

We have a new band! I will pass out "new arrival" candy cigars shortly. The new *60-meter band* is available for General Class operators and higher. We have been allocated 5 channels for upper sideband only transmissions, 2.8 kHz bandwidth, and effective radiated power of *50 watts or less*. The one thing that was NOT restricted was how high up our antenna can be, so 60 meters on my 100-foot tower is really going to play well! [97.303(s)] **ANSWER C**

G2D12 Which of the following is required by the FCC rules when operating in the 60 meter band?

A. If you are using other than a dipole antenna, you must keep a record of the gain of your antenna
B. You must keep a log of the date, time, frequency, power level and stations worked
C. You must keep a log of all third party traffic
D. You must keep a log of the manufacturer of your equipment and the antenna used

FCC rules require *60-meter band*, 5-channel operation not to exceed 50 watts effective radiated power out. On a dipole antenna, gain is zero, so 50 watts into the dipole from your transmitter will NOT exceed 50 watts effective radiated power output. However, if you're *transmitting into a home-made beam*, you will need to turn your power output down to correspond with the amount of gain the beam may exhibit. If the beam offers 3 dB gain in the forward direction, this 2 times increase in effective radiated power will require you to reduce your transmitter power output to -3 dB, or half of 50 watts. You then must note this in a *station logbook* and keep it as a permanent record as part of your station written files. [97.303(s)] **ANSWER A**

G2D07 Which of the following information must a licensee retain as part of their station records?
A. The call sign of other amateurs operating your station
B. Antenna gain calculations or manufacturer's data for antennas used on 60 meters
C. A record of all contacts made with stations in foreign countries
D. A copy of all third party messages sent through your station

The five channels on 60 meters are fun places to make contacts because you always know you are right on everyone's frequency. Moving slightly off frequency is absolutely not allowed on "60." On *60 meters*, effective radiated power shall be no more than 50 watts when compared to the unity gain of a dipole. You should *keep a written log of your antenna gain calculations*, and also note that you have decreased your transceiver's power output down to 50 watts into a unity gain dipole antenna. While log books are not normally required for routine contacts, a station log is always a handy reference guide. [97.103(c),(s)] **ANSWER B**

G1A16 Which of the following operating restrictions applies to amateur radio stations as a secondary service in the 60 meter band?
A. They must not cause harmful interference to stations operating in other radio services
B. They must transmit no more than 30 minutes during each hour to minimize harmful interference to other radio services
C. They must use lower sideband, suppressed-carrier, only
D. They must not exceed 2.0 kHz of bandwidth

Here we are, operating on one of the new 5 channels on 60 meters, and all of a sudden our conversation is bombarded by an incessant data signal coming from a nearby military facility. Since *we are secondary* for using the band, the military goes first, and we *must not cause interference* to them and expect that our comms will get clobbered now and then. [97.303(s)] **ANSWER A**

G1A15 What must you do if, when operating on either the 30 or 60 meter bands, a station in the primary service interferes with your contact?
A. Notify the FCC's regional Engineer in Charge of the interference
B. Increase your transmitter's power to overcome the interference
C. Attempt to contact the station and request that it stop the interference
D. Stop transmitting at once and/or move to a clear frequency

Both our 30 meter and 60 meter bands are shared, and authorized government users always have priority. *So change frequency or stop transmitting.* [97.303] **ANSWER D**

G1A14 Which of the following applies when the FCC rules designate the amateur service as a secondary user and another service as a primary user on a band?
A. Amateur stations must obtain permission from a primary service station before operating on a frequency assigned to that station
B. Amateur stations are allowed to use the frequency band only during emergencies
C. Amateur stations are allowed to use the frequency band only if they do not cause harmful interference to primary users
D. Amateur stations may only operate during specific hours of the day, while primary users are permitted 24 hour use of the band

Our relatively new 60 meter channelized ham band is shared with a few government agencies. They are primary on the band, which means *we must not cause harmful interference* to government users. [97.303] **ANSWER C**

G1A07 Which of the following frequencies is within the General Class portion of the 75 meter phone band?

A. 1875 kHz   C. 3900 kHz
B. 3750 kHz   D. 4005 kHz

The 75/80 meter ham band is a good one for short range daytime communications, and long range nighttime skywave signals. 80 meters refers to the bottom of the band for CW and data, and *75 meters* refers to the top of the band for phone. 300 divided by 75 = 4.000 MHz. To convert MHz to kHz, move the decimal point 3 places to the right. *3900 kHz* is within the phone privileges for General Class. [97.301(d)] **ANSWER C**

G1A09 Which of the following frequencies is within the General Class portion of the 80 meter band?

A. 1855 kHz   C. 3560 kHz
B. 2560 kHz   D. 3650 kHz

General Class CW and data privileges for the *80 meter* band extend from 3525 kHz through 3600 kHz. *3560 kHz* is a good spot for CW, RTTY, and DATA. [97.301(d)] **ANSWER C**

G1C06 What is the maximum transmitting power an amateur station may use on 1825 kHz?

A. 200 watts PEP output   C. 1200 watts PEP output
B. 1000 watts PEP output   D. 1500 watts PEP output

As a new General Class operator, you are permitted to run up to *1500 watts* of Peak Envelope Power on all bands other than 60 meters and 30 meters. However, always run the minimum power necessary. [97.313(b)] **ANSWER D**

G1A12 When a General Class licensee is not permitted to use the entire voice portion of a particular band, which portion of the voice segment is generally available to them?

A. The lower end
B. The upper end
C. The lower end on frequencies below 7.3 MHz and the upper end on frequencies above 14.150 MHz
D. The upper end on frequencies below 7.3 MHz and the lower end on frequencies above 14.150 MHz

Study our handy band plan charts and see where your privileges end and begin. *Voice privileges*, called "phone," are always at the *top of the bands*. (OK, 60 meters has its specific 5 channels of voice from top to bottom!) [97.301] **ANSWER B**

G1A13 (D) Which amateur band is shared with the Citizens Radio Service?

A. 10 meters   C. 12 meters
B. 11 meters   D. None

If you know a ham operator who has been licensed for more than 50 years, he will tell you war stories about how we lost our 11 meter ham band to Citizens Radio Service. *We now share NO bands with CB.* [97.303] **ANSWER D**

G2B07 What is a band plan?

A. A voluntary guideline for band use beyond the divisions established by the FCC
B. A guideline from the FCC for making amateur frequency band allocations
C. A guideline from the ITU for making amateur frequency band allocations
D. A plan devised by a club to best use a frequency band during a contest

The worldwide General Class bands have specific designated areas where different operating modes take place within the band. This is called a band plan, and there are different areas reserved within each band for specific types of emissions. It's important to know the band plan before transmitting on any Amateur Radio frequency. Just because you may have earned new privileges for a particular portion of the band does not necessarily mean you may operate any way

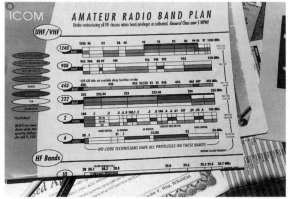

Band plans subdivide ham radio bands for specific uses, such as data, repeaters, and weak signal CW.

you want on that band. Certain segments of the worldwide band may be reserved for working foreign stations. There are other segments on the worldwide band for slow-scan television. There are still other segments for satellite reception.

You should do a lot of listening with your new General Class privileges, and only transmit to another SSB station when you hear them coming in loud and clear. This way you won't accidentally transmit voice on a portion of the band that might be reserved for, let's say, slow-scan television within the voice portion of the band. Remember, every band has a *band plan* and you must abide by the *voluntary guidelines*. **ANSWER A**

G2B08 What is the "DX window" in a voluntary band plan?

A. A portion of the band that should not be used for contacts between stations within the 48 contiguous United States
B. An FCC rule that prohibits contacts between stations within the United States and possessions on that band segment
C. An FCC rule that allows only digital contacts in that portion of the band
D. A portion of the band that has been set aside for digital contacts only

It is important to begin your General Class worldwide band operation in accordance with voluntary band plans. Most High Frequency bands have a spot where stateside hams will call and listen for *ONLY foreign country DX stations*. On the 160 meter band, 1830 kHz to 1850 kHz is long recognized as the DX WINDOW – no idle chit-chatting here! Use the High Frequency DX WINDOW as a great spot to listen for worldwide DX. **ANSWER A**

Elmer Point: *As a brand new General, spend a few days on the air LISTENING before you go on the air. Look at this band plan, and then tune in to hear how everything has its place on the radio dial. One of the best ways to complete your first transmission using your new General Class privileges is to answer an upbeat CQ call. Double-check to make sure you are within your General Class privileges, and check the band plan to make sure you are not answering a station looking only for foreign DX.*

160 Meters (1.8 – 2.0 MHz)

1.800 - 2.000	CW
1.800 - 1.810	Digital Modes
1.810	CW QRP
1.843 - 2.000	SSB, SSTV and other wideband modes
1.910	SSB QRP
1.995 - 2.000	Experimental
1.999 - 2.000	Beacons

80 Meters (3.5 – 4.0 MHz)

3.590	RTTY / Data DX
3.570 - 3.600	RTTY / Data
3,790 - 3.800	DX window
3.845	SSTV
3.885	AM calling frequency

40 Meters (7.0 – 7.3 MHz)

7.040	RTTY / Data DX
7.080 - 7.125	RTTY / Data
7.171	SSTV
7.290	AM calling frequency

30 Meters (10.1 – 10.15 MHz)

10.130 - 10.140	RTTY
10.140 - 10.150	Packet

20 Meters (14.0 – 14.35 MHz)

14.070 - 14.095	RTTY
14.095 - 14.0995	Packet
14.100	NCDXF Beacons
15,1005 - 14.112	Packet
14.230	SSTV
14.286	AM calling frequency

17 Meters (18.068 – 18.168 MHz)

18.100 - 18.105	RTTY
18.105 - 18.110	Packet

15 Meters (21.0 – 21.45 MHz)

21.070 - 21.110	RTTY / Data
21.340	SSTV

12 Meters (24.89 – 24.99 MHz)

24.920 - 24.925	RTTY
24.925 - 24.930	Packet

10 Meters (28.0 – 29.7 MHz)

28.000 - 28.070	CW
28.070 - 28.150	RTTY
28.150 - 28.190	CW
28.200 - 28.300	Beacons
28.300 - 29.300	Phone
28.680	SSTV
29.000 - 29.200	AM
29.300 - 29.510	Satellite Downliknks
29.520 - 29.590	Repeater Inputs
29.600	FM Simplex
29.610 - 29.700	Repeater Outputs

G2D08 Why do many amateurs keep a log even though the FCC doesn't require it?

A. The ITU requires a log of all international contacts

B. The ITU requires a log of all international third party traffic

C. The log provides evidence of operation needed to renew a license without retest

D. To help with a reply if the FCC requests information on who was control operator of your station at a given date and time

Whenever I let a guest ham talk over my equipment, I usually write down details of who the other operator was. This way, *if the FCC should ask* who was the control operator or third party during a transmission on a given date and time, I might be able to *look it up* and have a clue on what went out over the airwaves! **ANSWER D**

While the FCC doesn't require it, keeping a log of your station operation is a good idea.

G2D09 What information is traditionally contained in a station log?

A. Date and time of contact

B. Band and/or frequency of the contact

C. Call sign of station contacted and the signal report given

D. All of these choices are correct

I write down *all of my station operations* in the log book, and this way my notes are continuously available in case I need to go back and figure out why I got this QSL card from someone who said they worked me out on the boat. I usually write down my latitude and longitude while operating mobile marine, and this way I can send them a card back with all the details found in my log. **ANSWER D**

G2D01 What is the Amateur Auxiliary to the FCC?

A. Amateur volunteers who are formally enlisted to monitor the airwaves for rules violations

B. Amateur volunteers who conduct amateur licensing examinations

C. Amateur volunteers who conduct frequency coordination for amateur VHF repeaters

D. Amateur volunteers who use their station equipment to help civil defense organizations in times of emergency

The Amateur Auxiliary is made up of *volunteer hams* who are formally enlisted to *monitor the airwaves* for rule violations, and who report violations to the local FCC Compliance and Information Bureau office. **ANSWER A**

G2D02 What are the objectives of the Amateur Auxiliary?
A. To conduct efficient and orderly amateur licensing examinations
B. To encourage amateur self-regulation and compliance with the rules
C. To coordinate repeaters for efficient and orderly spectrum usage
D. To provide emergency and public safety communications

The benchmark of the amateur service is *self-regulation and compliance* with the rules. Hams help other hams stay on the straight and narrow. The FCC has little time nor budget to monitor the amateur bands for rule violations. **ANSWER B**

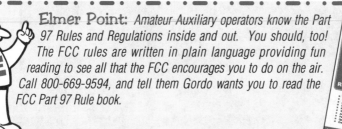

Elmer Point: *Amateur Auxiliary operators know the Part 97 Rules and Regulations inside and out. You should, too! The FCC rules are written in plain language providing fun reading to see all that the FCC encourages you to do on the air. Call 800-669-9594, and tell them Gordo wants you to read the FCC Part 97 Rule book.*

G2D03 What skills learned during "Fox Hunts" are of help to the Amateur Auxiliary?

A. Identification of out of band operation
B. Direction-finding skills used to locate stations violating FCC Rules
C. Identification of different call signs
D. Hunters have an opportunity to transmit on non-amateur frequencies

If you have been invited for a "fox hunt," forget about snares and nets, and bring your headphones and directional antenna. Amateur operators who are good at *signal direction-finding* can really help everyone pinpoint interference, locate stuck carriers, and even assist the Federal Communications Commission in tracking down jammers. **ANSWER B**

W6JAY practices his fox hunt direction-finding skills

G1B11 How does the FCC require an amateur station to be operated in all respects not covered by the Part 97 rules?
A. In conformance with the rules of the IARU
B. In conformance with amateur radio custom
C. In conformance with good engineering and good amateur practice
D. All of these answers are correct

When hams come up with new radio signaling techniques, they might not be found in the FCC rules. However, if this new signaling technique *conforms to good engineering and good amateur practice*, you can begin operating even though the rules may not specifically authorize this new particular type of radio emission. [97.101(a)] **ANSWER C**

G1B12 Who or what determines "good engineering and good amateur practice" that apply to operation of an amateur station in all respects not covered by the Part 97 rules?

A. The FCC

B. The Control Operator

C. The IEEE

D. The ITU

Only the Federal Communications Commission *(FCC)* can decide what meets "good engineering and good amateur practice." [97.101(a)] **ANSWER A**

Always follow good amateur and engineering practices when operating your ham station

---

G1B13 What restrictions may the FCC place on an amateur station that is causing interference to a broadcast receiver of good engineering design?

A. Restrict the amateur station operation to times other than 8 pm to 10:30 pm local time every day, as well as on Sundays from 10:30 am to 1 pm local time

B. Restrict the amateur station from operating at times requested by the owner of the receiver

C. Restrict the amateur station to operation only during RACES drills

D. Restrict the amateur station from operating at any time

As a new General Class operator, begin exploring your new privileges running the least amount of power possible. If you run full power and Granddad's older radio is tearing up FM music reception by your neighbor, the *FCC could impose "quiet hours"* where you could not transmit between *8 PM to 10:30 PM local time, daily,* as well as *Sundays from 10:30 AM to 1:00 PM* local time. Work with your neighbor to head off any confrontations. [97.121(a)] **ANSWER A**

G1B08 Which of the following is prohibited by the FCC Rules for amateur radio stations?

A. Transmission of music as the primary program material during a contact

B. The use of obscene or indecent words

C. Transmission of false or deceptive messages or signals

D. All of these answers are correct

Easy answer – hams *shall not play music* over the airwaves, nor may they *use obscene* and indecent *words,* or ever *transmit a false* or deceptive *message*. [97.113(a)(4), 97.113(e)] **ANSWER D**

G1B05 When may music be transmitted by an amateur station?

A. At any time, as long as it produces no spurious emissions
B. When it is unintentionally transmitted from the background at the transmitter
C. When it is transmitted on frequencies above 1215 MHz
D. When it is an incidental part of a space shuttle or ISS retransmission

About the only music you will ever hear on the ham bands is limited to some of the *audio feeds from the space shuttle or international space station retransmissions* where they sometimes wake up the crew with reveille, or sing "Happy Birthday" to them in outer space. All other forms of music are not allowed.
[97.113(a)(4),(e)] **ANSWER D**

Space Shuttle
Photo courtesy of N.A.S.A.

G1B06 When is an amateur station permitted to transmit secret codes?

A. During a declared communications emergency
B. To control a space station
C. Only when the information is of a routine, personal nature
D. Only with Special Temporary Authorization from the FCC

We encourage you to join AMSAT – Radio Amateur Satellite Corporation, a nonprofit scientific organization that supports our ham satellite programs. Specific AMSAT ham stations are allowed to *transmit secure secret codes to control space station* ham radio equipment. [97.113(a)(4) and 97.207(f)] **ANSWER B**

G1B07 What are the restrictions on the use of abbreviations or procedural signals in the amateur service?

A. Only "Q" codes are permitted

B. They may be used if they do not obscure the meaning of a message

C. They are not permitted because they obscure the meaning of a message to FCC monitoring stations

D. Only "10-codes" are permitted

Common abbreviations on the ham bands *do not obscure the meaning of our comms* and such phrases as "73," "QRZ?," or "Please QSL" are perfectly acceptable. Popular Q signals are given in the Appendix. The following is a list of common prowords. [97.113(a)(4)] **ANSWER B**

Proword	Meaning	Proword	Meaning
Affirmative	Yes	Number	Message number (in numerals) follows
All after	Say again all after _____	Out	End of transmission, no answer required or expected
All before	Say again all before _____		
Break	Used to separate message heading, text and ending	Over	End of transmission, answer is expected. Go ahead. Transmit.
Break	Stop transmitting		
Correct	That is correct	Roger	I have received your transmission satisfactorily
Figures	Numerals follow		
From	Originator follows	Say again	Repeat
Groups	Numeral(s) indicating numberof text words follows	Slant	Slant bar
		This is	This transmission is from the station whose call sign follows
Incorrect	That is incorrect		
Initial	Single letter follows	Time	File time or date-time group of the message follows
I say again	I repeat		
I spell	Phonetic spelling follows	To	Addressee follows
Message follows	A message which requires recording follows	Wait	Short pause
		Wait out	Long pause
More to follow	I have more traffic for you	Word after	Say again word after _____
Negative	No, not received	Word before	Say again word before _____

MARS – Army Radiotelephone Prowords

G1B09 When may an amateur station transmit communications in which the licensee or control operator has a pecuniary (monetary) interest?

A. Only when other amateurs are being notified of the sale of apparatus normally used in an amateur station and such activity is not done on a regular basis

B. Only when there is no other means of communications readily available

C. At any time as long as the communication does not involve a third party

D. Never

It is *okay to sell your ham equipment over the air* if is done on an occasional basis. [97.113(a)(3)] **ANSWER A**

G1E01 Which of the following would disqualify a third party from participating in stating a message over an amateur station?

A. The third party is a person previously licensed in the amateur service whose license had been revoked

B. The third party is not a U.S. citizen

C. The third party is a licensed amateur

D. The third party is speaking in a language other than English, French, or Spanish

You are *not allowed* to let a "third party" talk over your equipment if they were *previously licensed* as a ham and their *license has been revoked*. Don't let them "con" you into letting them speak over the microphone! [97.115(b)(2)]
ANSWER A

G1E05 What types of messages for a third party in another country may be transmitted by an amateur station?

A. Any message, as long as the amateur operator is not paid

B. Only messages for other licensed amateurs

C. Only messages relating to amateur radio or remarks of a personal character, or messages relating to emergencies or disaster relief

D. No messages may be transmitted to foreign countries for third parties

During third party traffic exchange, make sure that the station in the other country only transmits *personal messages, or*, in an *emergency*, they are ok to transmit messages relating to disaster relief. [97.115(a)(2), 97.117] **ANSWER C**

Elmer Point: *Before you allow third party traffic at you radio station, make sure your guest operator understands the rules: no business; no profanity; no secret codes; no music, and only a language that you can understand. It's also a good idea to keep a logbook with details of the third party conversation.*

List of Countries Permitting Third-Party Traffic

Country	Call Sign Prefix	Country	Call Sign Prefix	Country	Call Sign Prefix
Antigua and Barbuda	V2	El Salvador	YS	Paraguay	ZP
Argentina	LU	The Gambia	C5	Peru	OA
Australia	VK	Ghana	9G	Philippines	DU
Austria, Vienna	4U1VIC	Grenada	J3	Pitcairn Island	VR6
Belize	V3	Guatemala	TG	St. Christopher & Nevis	V4
Bolivia	CP	Guyana	8R	St. Lucia	J6
Bosnia-Herzegovina	T9	Haiti	HH	St. Vincent & Grenadines	J8
Brazil	PY	Honduras	HR	Sierra Leone	9L
Canada	VE, VO, VY	Israel	4X	South Africa	ZS
Chile	CE	Jamaica	6Y	Swaziland	3D6
Colombia	HK	Jordan	JY	Trinidad and Tobago	9Y
Comoros	D6	Liberia	EL	Turkey	TA
Costa Rica	TI	Marshall Is	V6	United Kingdom	GB
Cuba	CO	Mexico	XE	Uruguay	CX
Dominica	J7	Micronesia	V6	Venezuela	YV
Dominican Republic	HI	Nicaragua	YN	ITU-Geneva	4U1ITU
Ecuador	HC	Panama	HP	VIC-Vienna	4U1VIC

G1E10 Which of the following is a permissible third party communication during routine amateur radio operations?
- A. Permitting an unlicensed person to speak to a licensed amateur anywhere in the world
- B. Sending a business message for another person, as long it is for a non-profit organization
- C. Sending a business message for another person, as long as the control operator has no pecuniary interest in the message
- D. Sending a message to a third party through a foreign station as long as that person is a licensed amateur radio operator

We pass third party traffic from one ham to another, so, as long as that *person in* the *other country is an amateur radio operator* and we have a third party agreement with that country, we are good to go! [97.115(a)(2)] **ANSWER D**

G1E07 With which of the following is third-party traffic prohibited, except for messages directly involving emergencies or disaster relief communications?
- A. Countries in ITU Region 2
- B. Countries in ITU Region 1
- C. Any country other than the United States, unless there is a third-party agreement in effect with that country
- D. Any country which is not a member of the Internal Amateur Radio Union (IARU)

This question asks where third party traffic is PROHIBITED. Except in an emergency, *we are prohibited* form passing international third party traffic *unless we have a third party agreement* in effect with that country. [97.115(a)(2)]
ANSWER C

G1E08 Which of the following is a requirement for a non-licensed person to communicate with a foreign amateur radio station from a US amateur station at which a licensed control operator is present?
- A. Information must be exchanged in English
- B. The foreign amateur station must be in a country with which the United States has a third party agreement
- C. The control operator must have at least a General Class license
- D. All of these answers are correct

To pass third party traffic to a station in another country, we must have a *third party agreement.* [97.115(a)(b)] **ANSWER B**

The following CEPT countries allow U.S. Amateurs to operate in their countries without a reciprocal license. Be sure to carry a copy of your FCC license and FCC Public Notice DA99-1098.

Austria	Finland	Liechtenstein	Slovenia
Belgium	France & its	Lithuania	Spain
Bosnia & Herzegovina	possessions	Luxembourg	Sweden
Bulgaria	Germany	Monaco	Switzerland
Croatia	Greenland	Netherlands	Turkey
Cyprus	Hungary	Netherlands Antilles	United Kingdom & its
Czech Republic	Iceland	Norway	possessions
Denmark	Ireland	Portugal	
Estonia	Italy	Romania	
Faroe Islands	Latvia	Slovak Republic	

G1D13 When may you participate as a VE in administering an amateur radio license examination?
- A. Once you have notified the FCC that you want to give an examination
- B. Once you have a Certificate of Successful Completion of Examination (CSCE) for General Class
- C. Once your General Class license appears in the FCC's ULS database
- D. Once you have been granted your General Class license and received your VEC accreditation

It will take about 10 days to see your *General Class license* grant on the FCC's ULS site. This is proof positive that you can now ask a Contact Volunteer Examiner for the VE application. When completed and signed by your contact VE, your application is sent on to the *VEC for accreditation*. [97.509] **ANSWER D**

G1D02 What license examinations may you administer when you are an accredited VE holding a General Class operator license?

A. Novice	C. Technician
B. General	D. All elements

As a new General Class operator, I hope you will apply for Volunteer Examiner accreditation. You may need a Contact Volunteer Examiner to sponsor you. As a General Class operator, you and 2 other VE-certified General Class operators could administer an *Element 2 Technician Class* written examination. [97.509(b)(3)(i)] **ANSWER C**

G1D05 Which of the following is sufficient for you to be an administering VE for a Technician Class operator license examination?
- A. Notification to the FCC that you want to give an examination
- B. Receipt of a CSCE for General Class
- C. Possession of properly obtained telegraphy and written examinations
- D. A FCC General Class or higher license and VEC accreditation

To take part in an examination session, you will need to hold a minimum of an FCC *General Class* amateur license, and you must be *accredited by* a volunteer examiner coordinator *(VEC)*. And remember, it takes a minimum of 3 accredited examiners to be present to conduct the test session. We hope you will soon become an accredited examiner because we will need more exam givers for the Element 2 Technician written exam. [97.509(b)(3)(i)] **ANSWER D**

G1D12 Volunteer Examiners are accredited by what organization?
A. The Federal Communications Commission
B. The Universal Licensing System
C. A Volunteer Examiner Coordinator
D. The Wireless Telecommunications Bureau

Individual Volunteer Examiners are *accredited by a Volunteer Examiner Coordinator*. The largest VECs are the American Radio Relay League (ARRL) and the W5YI-VEC. [97.509(b)(1)] **ANSWER C**

Elmer Point: *Our listing of Volunteer Exam Coordinators will lead you to a phone number of your local area testing team. Look for it in the Appendix on page 200. Testing teams are always looking for examinees, so they'll be delighted to hear that you want to take an exam. And once you're a General, you may want to join them as an examiner!*

G1D04 Which of the following are requirements for administering a Technician Class operator examination?
A. At Least three VEC-accredited General Class or higher VEs must be present
B. At least two VEC-accredited General Class or higher VEs must be present
C. At least two General Class or higher VEs must be present, but only one need be VEC accredited
D. At least three VEs of Technician Class or higher must be present

As a General Class operator, you have capabilities to take part in the volunteer examination system. A contact VE may need to sign off on your VE application. That application goes to the Volunteer Exam Coordinator. *Once* you are *approved, you and 2 other accredited General Class* and higher volunteer examiners may administer a Technician Class (Element 2) operator examination. [97.509(a)(b)] **ANSWER A**

G1D11 What criteria must be met for a non U.S. citizen to be an accredited Volunteer Examiner?
A. The person must be a resident of the U.S. for a minimum of 5 years
B. The person must hold a U.S. amateur radio license of General Class or above
C. The person's home citizenship must be in the ITU 2 region
D. None of these answers is correct; non U.S. citizens cannot be volunteer examiners

Non U.S. citizens may become both ham operators with a U.S.A. license, as well as volunteer examiners if they *hold a U.S. General Class license or higher*. [97.509(b)(3)] **ANSWER B**

G1D10 What is the minimum age that one must be to qualify as an accredited Volunteer Examiner?

    A. 12 years   C. 21 years

    B. 18 years   D. There is no age limit

Eighteen years of age *or older* for volunteer examiner accreditation. Remember, while there is no age limit to obtain the ham license, volunteer examiners must be at least 18 years or older. [97.509(b)(2)] **ANSWER B**

Kids make great ham radio operators, but you have to be 18 or older to become a VE.

G1D07 Who is responsible at a Volunteer Exam Session for determining the correctness of the answers on the exam?

    A. The FCC   C. The VEC

    B. The administering VEs   D. The local VE team liaison

I hope you will become a volunteer examiner. We need more volunteer examiners as our ranks swell with new hams getting to General Class without the "dreaded" Morse code test. Occasionally you may be testing a disabled applicant, and it is up to the *administering volunteer examiners* how they wish to conduct the exam and how they may *judge the correctness of the answers*. [97.509(h)] **ANSWER B**

Elmer Point: *"Roger on your QTH, and fine business on your new rig. Hope you can QSL our QSO, and I wish you very seven three." Say what? Huh? Here's a glossary with a sampling of ham radio lingo that you'll hear when you're on the air!*

CQ CQ CQ this is a station on HF looking for anyone to have a nice friendly contact

CQDX CQDX CQDX this is a station calling ONLY for a contact to a foreign station, not looking for a USA contact

Old Man this really doesn't mean you are old, but rather a term for a fellow ham radio operator

YL usually an unmarried lady

XYL usually a married lady

Harmonics your children

73 best regards to you

88 an affectionate hug to a YL or XYL

Ham a licensed amateur radio operator (with every ham having his or her own idea of this term's origin)

ARRL American Radio Relay League, our number one, non-profit organization that you should join.

APRS Automatic Position Reporting System - automatic GPS radio tracking, usually on 10 MHz HF

DX a station a LONG way away

Break only use this word to break into a conversation with emergency or priority traffic.
To enter a conversation, politely, only by using your callsign, never "Break".

CQ contest you can answer this call if you are familiar with the precise "exchange" the other operator is looking for.

QRZed who is the station calling me?

Down 10 move down in frequency 10 kHz

5 9 9 your signal report is strong, loud, and clear!

In the Mud your signal is extremely weak

QRM interference from another station on a close frequency

QTH your station location

QSL please send me a QSL card   (also...I agree)

QRN power line or automobile static

QRU does anyone have traffic for me?

QRP low power station

QRT going off the air

QRX stand by - I need to do something

QSV your signal is fading in and out

QSO a communications contact

QST calling all ham radio operators

QSY we need to move off this frequency

TVI interfering with a television set

Heil a premier after-market microphone system

HF worldwide high frequency bands

MARS Military Affiliate Radio Service

Nets regular frequency meeting spot at a certain time for all interested hams

OO an official observer monitoring for rule violations

My Shack your home radio location

Mobile in motion driving down the road with a big HF rig

Ragchew having a long winded conversation

RFI every time I transmit, my windshield wipers self-activate!

Handi-ham an active radio ham who has overcome physical challenges

Splatter maybe turn down your microphone gain

Traffic an incoming message for you

HI HI radio laughter on CW

G2B12 What is a practical way to avoid harmful interference when selecting a frequency to call CQ using phone?
- A. Ask if the frequency is in use, say your call sign, and listen for a response
- B. Keep your CQ to less than 2 minutes in length to avoid interference to contacts that may be in progress
- C. Listen for 2 minutes before calling CQ to avoid interference to contacts that may be in progress
- D. Call CQ at low power first and if there is no indication of interference, increase power as necessary

On the worldwide ham bands with your new General Class privileges, first monitor a quiet frequency for about one minute before transmitting. Then, key the mike and *say, "Is the frequency in use?" followed by your call sign*. Then release the PTT (push-to-talk button) and see if anyone else is using the frequency. **ANSWER A**

G2A12 What is the recommended way to break into a conversation when using phone?
- A. Say "QRZ" several times followed by your call sign
- B. Say your call sign during a break between transmissions from the other stations
- C. Say "Break" "Break" "Break" and wait for a response
- D. Say "CQ" followed by the call sign of either station

When you hear a conversation between two hams, a polite way to join this QSO (communications) is to simply *say your call sign in between* one station turning it over to the other station. Say your call sign in a cheerful way, making it sound like you wish to enter the conversation in a friendly way. Don't just blurt out your call sign – sound pleasant, as if asking permission to join in on the QSO. **ANSWER B**

G2A13 What does the expression "CQ DX" usually indicate?
- A. A general call for any station
- B. The caller is listening for a station in Germany
- C. The caller is looking for any station outside their own country
- D. This is a form of distress call

The term "CQ" is used by hams to "fish" for a new station to answer their call. Say your "CQ" with a smile and sound excited about making the call to anyone hearing you. A drab, lifeless "CQ" is like fishing with old bait. If you hear someone calling *"CQ DX"*, this means they are not looking for just any stateside contact, but rather they are *seeking calls ONLY from very distant or foreign stations* thousands of miles away. These usually are very experienced operators, so stay tuned and learn from the "pros" how to call out and get rare "DX" responses. **ANSWER C**

G2B01 What action should be taken if the frequency on which a net normally meets is in use just before the net begins?
    A. Reduce your output power and start the net as usual
    B. Increase your power output so that net participants will be able to hear you
    C. Ask the stations if the net may use the frequency, or move the net to a nearby clear frequency if necessary
    D. Cancel the net for that day

As a new General Class operator, you are going to find many daytime and evening nets on the worldwide frequencies conducted on a daily basis. Every so often the frequency is in use before the net begins. Courteous net controllers may *ask the operating stations whether or not the net can go on this same frequency* while the two stations previously operating stand by. An *alternate* method would be to *conduct the net on a nearby frequency*, no closer than 3- to 5-kHz away from the regular net frequency. **ANSWER C**

G2B02 What should be done if a net is about to begin on a frequency you and another station are using?
    A. Move to a different frequency as a courtesy to the net
    B. Tell the net that they must to move to another frequency
    C. Reduce power to avoid interfering with the net
    D. Pause between transmissions to give the net a chance to change frequency

If you are asked to move prior to a net beginning, as a *courtesy* to all ham operators, *shift off the frequency*. **ANSWER A**

G2B03 What should you do if you notice increasing interference from other activity on a frequency you are using?
    A. Tell the interfering stations to change frequency since you were there first
    B. Report the interference to your local Amateur Auxiliary Coordinator
    C. Move your contact to another frequency
    D. Turn on your amplifier

On the worldwide General Class ham bands, propagation will sometimes cause stations on the same frequency to all of a sudden come in right on top of your ongoing contact. Be a good ham and *move your contact to another frequency*, if you can, to avoid the interference. **ANSWER C**

G2B05 What minimum frequency separation between SSB signals should be allowed to minimize interference?
    A. Between 150 and 500 Hz   C. Approximately 6 kHz
    B. Approximately 3 kHz   D. Approximately 10 kHz

When operating single sideband, your emission will take up *approximately 3 kHz* of bandwidth. Always stay at least 3 kHz away from any other station on the air that is using an adjacent frequency. **ANSWER B**

G2D10 What is QRP operation?
    A. Remote Piloted Model control
    B. Low power transmit operation, typically about 5 watts
    C. Transmission using Quick Response Protocol
    D. Traffic Relay Procedure net operation

Many ham operators on worldwide frequencies enjoy operating *QRP* – the Q code for *low power operation*. Hams get a big kick out of working all the way around the world with less power than that required by a tiny flashlight light bulb. You can look for QRP operation at specific spots on the radio dial band plan. **ANSWER B**

G2A10 Which of the following statements is true of VOX operation?

A. The received signal is more natural sounding
B. VOX allows "hands free" operation
C. Frequency spectrum is conserved
D. The duty cycle of the transmitter is reduced

Most worldwide radios have a voice-operated relay circuit, abbreviated *"VOX," for hands-free operation.* If you have a big base station microphone, this is a neat circuit to minimize you having to reach over to depress the push-to-talk switch. Just be sure you never leave the VOX circuit turned on when you are not right at your turned-on equipment! **ANSWER B**

Using a headset with an attached mike on VOX will keep both hands free when you're taking on the worldwide bands while at your home station.

G2A11 Which of the following user adjustable controls are usually associated with VOX circuitry?

A. Anti-VOX
B. VOX Delay
C. VOX Sensitivity
D. All of these choices are correct

Anti-VOX keeps the output from your radio's speaker from tripping its own transmit VOX circuit. *VOX delay* keeps the equipment keyed up during syllables, and is usually pre-set to the point where normal conversation doesn't cause your equipment to chatter in and out of transmit. *VOX sensitivity* is set so that normal breathing, and an occasional sneeze, WON'T trigger your set into transmit.
ANSWER D

G1E04 Which of the following conditions require an amateur radio station to take specific steps to avoid harmful interference to other users or facilities?

A. When operating within one mile of an FCC Monitoring Station
B. When using a band where the amateur service is secondary
C. When a station is transmitting spread spectrum emissions
D. All of these answers are correct

A relatively new form of ham radio emission is called "spread spectrum" and you must take specific steps to avoid harmful interference to other stations. Also, when operating on 30 meters and 60 meters on High Frequency, we are secondary to the frequencies, and must not cause interference. Also, if you live within one mile of an FCC monitoring station, take steps to avoid any harmful interference. You do NOT want the FCC knocking on your door. *All of these answers are correct.*
[97.13(b), 97.311(b), 97.303] **ANSWER D**

G1E03 What kind of amateur station simultaneously retransmits the signals of other stations on another channel?

A. Repeater Station
B. Beacon Station
C. Telecommand Station
D. Relay Station

An amateur station that automatically and simultaneously retransmits other amateur stations is called a *repeater station*. Watch out for Answer D – the official name is a "repeater." [97.3(a)(39)] **ANSWER A**

G1E06 Which of the following applies in the event of interference between a coordinated repeater and an uncoordinated repeater?
A. The licensee of the non-coordinated repeater has primary responsibility to resolve the interference
B. The licensee of the coordinated repeater has primary responsibility to resolve the interference
C. Both repeater licensees share equal responsibility to resolve the interference
D. The frequency coordinator bears primary responsibility to resolve the interference

If you decide to put up your own repeater, better have a big bank account! They are expensive. Also, seek coordination, because a *non-coordinated repeater has primary responsibility* to resolve interference to another repeater on the same frequency. [97.205(c)] **ANSWER A**

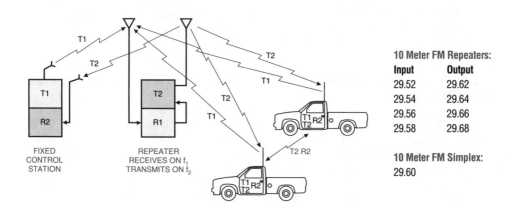

10 Meter FM Repeaters:	
Input	**Output**
29.52	29.62
29.54	29.64
29.56	29.66
29.58	29.68

10 Meter FM Simplex:
29.60

Repeater
Source: *Mobile 2-Way Radio Communications*, G. West, Copyright ©1992 Master Publishing, Inc., Niles, IL

G4E01 Which of the following emission types are permissible while operating HF mobile?
A. CW
B. SSB
C. FM
D. All of these choices are correct

While operating high frequency mobile, *you are permitted all emission modes* that are allowed for the particular band of operation. This includes Morse code mobile! But don't try operating digital modes while looking at the screen as you are driving down the street! **ANSWER D**

G8B05 Why isn't frequency modulated (FM) phone used below 29.5 MHz?
A. The transmitter efficiency for this mode is low
B. Harmonics could not be attenuated to practical levels
C. The bandwidth would exceed FCC limits
D. The frequency stability would not be adequate

We cannot use frequency modulation phone below 29.5 MHz because the *bandwidth is simply too wide*. Most FM simplex operation is found on 29.6 MHz, a good spot to enjoy your new General Class privileges. There are also repeaters up in this range. **ANSWER C**

G2A05 Which mode of voice communication is most commonly used on the High Frequency Amateur bands?

A. FM   C. SSB

B. AM   D. PM

When you gain your new privileges on High Frequency, the majority of voice modes are always *single sideband, SSB*. **ANSWER C**

G2A06 Which of the following is an advantage when using single sideband as compared to other voice modes on the HF amateur bands?

A. Very high fidelity voice modulation

B. Less bandwidth used and high power efficiency

C. Ease of tuning on receive

D. Less subject to static crashes (atmospherics)

The advantage of *single sideband* is it *occupies less spectrum* than double-sideband AM, is power efficient, and there is no continuous carrier in between syllables in your transmission. **ANSWER B**

G2A08 Which of the following statements is true of single sideband (SSB) voice mode?

A. It is a form of amplitude modulation in which one sideband and the carrier are suppressed

B. It is a form of frequency modulation in which higher frequencies are emphasized

C. It reproduces upper frequencies more efficiently than lower frequencies

D. It is the only voice mode authorized on the HF bands between 14 and 30 MHz

When you first start using your new SSB transceiver, turn the big knob rapidly when you begin to hear the sounds of a voice signal coming through. Bring the voice into a natural sounding signal by rotating the big knob to clear up that "Donald Duck" sound! *Single sideband is a form of AM modulation.* **ANSWER A**

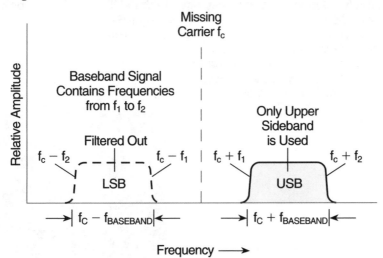

SSB signals are Amplitude Modulated (AM)
with the carrier and one sideband suppressed.

G2A07 Which of the following statements is true of the single sideband (SSB) voice mode?

A. Only one sideband and the carrier are transmitted; the other sideband is suppressed

B. Only one sideband is transmitted; the other sideband and carrier are suppressed

C. SSB voice transmissions have higher average power than any other mode

D. SSB is the only mode that is authorized on the 160, 75 and 40 meter amateur bands

When we transmit using *SSB, only a single sideband*, approximately 2.8 kHz wide, *is sent out over the air*. The opposite sideband is suppressed. There also is no carrier in a properly adjusted SSB signal. This means your radio gets a complete rest during each break in your syllables! This is great for battery operation in the field. **ANSWER B**

G2A04 Which mode is most commonly used for voice communications on the 17 and 12 meter bands?

A. Upper Sideband   C. Vestigial Sideband

B. Lower Sideband   D. Double Sideband

Since 17- and 12-meters are higher in frequency than 20 meters, we always use *upper sideband*. **ANSWER A**

G2A01 Which sideband is most commonly used for phone communications on the bands above 20 meters?

A. Upper Sideband   C. Vestigial Sideband

B. Lower Sideband   D. Double Sideband

We normally use *upper sideband (USB)* for all voice emissions on 20-, 17-, 15-, 12-, and 10-meters, the 5 channels on our new 60-meter band, and on all of the VHF and UHF weak-signal portions of the bands. **ANSWER A**

G2A03 Which sideband is commonly used in the VHF and UHF bands?

A. Upper Sideband   C. Vestigial Sideband

B. Lower Sideband   D. Double Sideband

When you upgrade to General Class, we hope you will continue to stay active on VHF and UHF bands, too. If you operate weak signal on VHF and UHF, our mode is always *upper sideband*. **ANSWER A**

When you're using voice — even when you're operating portable on the beach — make sure to use the correct sideband.

G2A02 Which sideband is commonly used on the 160, 75, and 40 meter bands?

A. Upper Sideband
B. Lower Sideband
C. Vestigial Sideband
D. Double Sideband

We use *lower sideband* (LSB) on 160-, 80-, and 40-meters. And while it is not absolutely illegal to use lower sideband on 20 meters and shorter wavelength bands, good operating procedure would always indicate "go with the flow" and use the proper sideband. **ANSWER B**

G2A09 Why do most amateur stations use lower sideband on the 160, 75 and 40 meter bands?

A. The lower sideband is more efficient at these frequency bands
B. The lower sideband is the only sideband legal on these frequency bands
C. Because it is fully compatible with an AM detector
D. Current amateur practice is to use lower sideband on these frequency bands

Remember that the choice of upper and lower sideband is by gentleman's agreement, and you won't find it in the FCC rule book. On the 160-, 75-, and 40-meter bands, *amateur "practice" is to use lower sideband*. Most modern amateur worldwide high-frequency equipment automatically selects upper sideband for the bands 20 meters and up, and lower sideband for the bands 40 meters and down. **ANSWER D**

SIDEBAND	FREQUENCY BAND IN METERS								
USB			60		20	17	15	12	10	
LSB	160	75/80		40						

Sideband Usage on Amateur Radio Bands

Elmer Point: *Want to learn more about how radios work? I suggest you read and study* Basic Communications Electronics *by Jack Hudson, W9MU, and Jerry Luecke, KB5TZY. It explains how transmitters and receivers work, the science behind antennas, integrated circuits, and more. You can pick up a copy at your ham radio store, online at www.w5yi.com, or by calling W5YI Group at 800-669-9594. Understanding how radios work will add to your enjoyment of your General Class privileges!*

G2B13 What is a practical way to avoid harmful interference when calling CQ using Morse code or CW?
A. Send the letter "V" 12 times and then listen for a response
B. Keep your CQ to less than 2 minutes in length to avoid interference with contacts already in progress
C. Send "QRL? de" followed by your call sign and listen for a response
D. Call CQ at low power first; if there is no indication of interference then increase power as necessary

When operating on CW, before transmitting a CQ, first listen for about one minute to ensure the frequency is clear, and then *send "QRL?" followed by your call sign*. The Q-code "QRL" means "Are you busy – is this frequency occupied?" Some CW operators also send just the question mark to check and see whether or not the frequency is available. **ANSWER C**

G2B04 What minimum frequency separation between CW signals should be allowed to minimize interference?
A. 5 to 50 Hz
B. 150 to 500 Hz
C. 1 to 3 kHz
D. 3 to 6 kHz

When operating CW, try to separate yourself from other CW transmissions by at least *150 Hz to 500 Hz*. This will give your CW signal a distinct tonal difference from the other station, and hopefully will not cause interference. **ANSWER B**

G2F06 What does the term "zero beat" mean in CW operation?
A. Matching the speed of the transmitting station
B. Operating split to avoid interference on frequency
C. Sending without error
D. Matching the frequency of the transmitting station

When a net control station asks everyone to *"zero beat"* their CW operation, they would like you to *match their frequency* so that the pitch of everyone's CW tone will sound about the same. **ANSWER D**

G2F05 What is the best speed to use answering a CQ in Morse code?
A. The speed at which you are most comfortable copying
B. The speed at which the CQ was sent
C. A slow speed until contact is established
D. 5 wpm, as all operators licensed to operate CW can copy this speed

Ready to try your first CW CQ? *Send at a relatively slow rate*, and expect that any other station will send at this *slow rate in their response* to your CQ. Same thing applies in reverse. If you hear a station sending CQ at a ridiculously slow speed, chances are they are brand new on Morse code, so send back to them at the exaggerated slower speed. **ANSWER B**

G2FO7 When sending CW, what does a "C" mean when added to the RST report?
A. Chirpy or unstable signal
B. Report was read from S meter reading rather than estimated
C. 100 percent copy
D. Key clicks

If someone sends you an RST report of 5-9-9 *C*, it means your signal is unstable or *chirping*, probably due to an inadequate power supply or operating from a low battery. **ANSWER A**

Elmer Point: *The S meter on the front of your radio indicates signal strength. S-9 is much stronger than S-5, and S-1 is relatively weak, but it will be up to your own ears and brain to judge readability. R-2 means you can make out the signal with difficulty. R-5 is a loud and clear signal report. A great report would be 5 by 9. A bad one might be 3 by 3. Sometimes hams will generalize for strength, readability and CW tone as Q-5. Let's hope you always get a 5 by 9! Here's the how the RST Signal Reporting System works:*

The RST system is a way of reporting on the quality of a received signal by using a three digit number. The first digit indicates Readability (R), the second digit indicates received Signal Strength (S), and the third digit indicates Tone (T).

READABILITY (R) for Voice + CW
1 – Unreadable
2 – Barely readable, occasional words distinguishable
3 – Readable with considerable difficulty
4 – Readable with practically no difficulty
5 – Perfectly readable

SIGNAL STRENGTH (S) for Voice + CW
1 – Faint, barely perceptible signals
2 – Very weak signals
3 – Weak signals
4 – Fair signals
5 – Fairly good signals
6 – Good signals
7 – Moderately strong signals
8 – Strong signals
9 – Extremely strong signals

TONE* (T) Use on CW only
1 – Very rough, broad signals, 60 cycle AC may be present
2 – Very rough AC tone, harsh, broad
3 – Rough, low-pitched AC tone, some trace of filtering
5 – Filtered, rectified AC note, musical, ripple modulated
6 – Slight trace of filtered tone but with ripple modulation
7 – Near DC tone but trace of ripple modulation
8 – Good DC tone, may have slight trace of modulation
9 – Purest, perfect DC tone with no trace of ripple or modulation

*The TONE report refers only to the purity of the signal, and has no connection with its stability or freedom from clicks or chirps. If the signal has the characteristic steadiness of crystal control, add X to the report (e.g., RST 469X). If it has a chirp or "tail" (either on "make" or "break") add C (e.g., RST 469C). If it has clicks or other noticeable keying transients, add K (e.g., 469K). If a signal has both chirps and clicks, add both C and K (e.g., 469CK).

G2F02 What should you do if a CW station sends "QRS" when using Morse code?
A. Send slower
B. Change frequency
C. Increase your power
D. Repeat everything twice

If you are sending code to another station, and they respond *QRS*, this means for you to please *send at a slower rate*. **ANSWER A**

G2F10 What does the Q signal "QRQ" mean when operating CW?
A. Slow down
C. Zero beat my signal
B. Send faster
D. Quitting operation
If you are sending CW to another station, and they send back *"QRQ"*, they wish
you to *send faster* because they are good at copying code at a higher speed.
ANSWER B

Elmer Point: *Q-R-Zed the station calling? I missed your call sign in the QRN. The other operator is saying please repeat your call sign because he didn't hear it clearly through the local noise. You don't need to memorize the Q codes but they are heard frequently in a conversation. Although the Q codes were developed for shortening Morse code messages, they also are popular using voice. A list of Q codes and common CW abbreviations appears in the Appendix on pages 206 and 207.*

G2F11 What does the Q signal "QRV" mean when operating CW?
A. You are sending too fast
B. There is interference on the frequency
C. I am quitting for the day
D. I am ready to receive messages
The Q code *"QRV"* means that you are *ready*, with pen or computer at hand,
to receive the incoming message. If you send "QRV?," you are asking, "are there
any messages holding for my station?" **ANSWER D**

G2F03 What does it mean when a CW operator sends "KN" at the end of a transmission?
A. Listening for novice stations
B. Operating full break-in
C. Listening only for a specific station or stations
D. Closing station now
When a station turns the communication back to you, and sends *"KN"* at the end of
their transmission, it means that they're wishing for *only you to respond* and all
other stations to stand by. **ANSWER C**

G2F08 What prosign is sent using CW to indicate the end of a formal message?
A. SK
C. AR
B. BK
D. KN
The *end* of a *formal CW message* usually includes *"AR."* This lets all operators
know that the formal message has been completely sent. **ANSWER C**

G2F09 What does the Q signal "QSL" mean when operating CW?
A. Send slower
B. We have already confirmed by card
C. I acknowledge receipt
D. We have worked before
The term *"QSL"* has several meanings in ham radio. On CW and sometimes
voice, it means that the other *station has acknowledged your message*. And if
the other station asks for a QSL, they are likely asking for you to send them a

colorful QSL card with your call sign on the front and QSO details on the back. QSL cards are fun to collect and all General Class hams should have their own QSL card. Also, you can QSL a contact electronically at several QSL e-mail sites. If ever you receive a QSL card, you must always send one back! **ANSWER C**

G2F04 What does it mean when a CW operator sends "CL" at the end of a transmission?
A. Keep frequency clear
B. Operating full break-in
C. Listening only for a specific station or stations
D. Closing station
When an operator sends *"CL"* or "SK" at the end of their transmission, it means that they are *closing down their station* and going off the air. **ANSWER D**

G2F01 Which of the following describes full break-in telegraphy (QSK)?
A. Breaking stations send the Morse code prosign BK
B. Automatic keyers are used to send Morse code instead of hand keys
C. An operator must activate a manual send/receive switch before and after every transmission
D. Incoming signals are received between transmitted code character elements
Although the FCC has eliminated an examination for knowing Morse code, learning the code is important to be a good ham. All high frequency transceivers have Morse code capabilities, and many also include an automatic, built-in code keyer. Just add paddles! You will start out using the VOX mode to control the transmitter to turn on when you start keying, and turn off after a few seconds of not-keying. This is called semi-break-in. And when you really get good at CW, you can set your transceiver to operate CW in the *QSK full break-in mode*. In between each dot and dash, the receiver will instantly tune in on what's happening as you are sending CW on the bands. If another station wishes to interrupt, you will *hear their signal between your dots and dashes*. **ANSWER D**

G1B03 Which of the following is a purpose of a beacon station as identified in the FCC Rules?
A. Observation of propagation and reception, or other related activities
B. Automatic Identification of Repeaters
C. Transmission of bulletins of General interest to amateur radio licensees
D. Identifying Net Frequencies
Beacon stations are important for the *study of propagation* from the ionosphere as well as the atmosphere. Always try to stay clear of beacon stations when selecting a frequency on which to transmit. Beacon stations are found at 14.10 MHz, 28.2 - 28.3 MHz, and up on the 2-meter band below 144.3 MHz. These are one-way transmissions. [97.1(a)(9)] **ANSWER A**

G1B10 What is the power limit for beacon stations?
A. 10 watts PEP output   C. 100 watts PEP output
B. 20 watts PEP output   D. 200 watts PEP output
Beacon stations used for propagation surveys must never transmit more than *100 watts* peak envelope power output. [97.203(c)] **ANSWER C**

G1B02 With which of the following conditions must beacon stations comply?
A. Identification must be in Morse code
B. The frequency must be coordinated with the National Beacon Organization
C. The frequency must be posted on the Internet or published in a national periodical
D. There must be no more than one beacon signal in the same band from a single location

Beacon stations are important for the study of propagation from the Ionosphere as well as the atmosphere. Do not transmit in beacon band plan frequencies. Ham operators are permitted to put up *only a single beacon signal*, in the same band, *from a single location*. [97.203(b)] **ANSWER D**

Radio Beacon Stations

Slot	Country	Call	14.100	18.110	21.150	24.930	28.200	Operator
1	United Nations	4U1UN	00:00	00:10	00:20	00:30	00:40	UNRC
2	Canada	VE8AT	00:10	00:20	00:30	00:40	00:50	RAC
3	USA	W6WX	00:20	00:30	00:40	00:50	01:00	NCDXF
4	Hawaii	KH6WO	00:30	00:40	00:50	01:00	01:10	UHRO
5	New Zealand	ZL	00:40	00:50	01:00	01:10	01:20	NZART
6	Australia	VK8	00:50	01:00	01:10	01:20	01:30	W1A
7	Japan	JA21CY	01:00	01:10	01:20	01:30	01:40	JARL
8	China	BY	01:10	01:20	01:30	01:40	01:50	CRSA
9	Russia	UA	01:20	01:30	01:40	01:50	02:00	TBO
10	Sri Lanka	4S7B	01:30	01:40	01:50	02:00	02:10	RSSL
11	South Africa	ZS6DN	01:40	01:50	02:00	02:10	02:20	ZS6DN
12	Kenya	5Z4B	01:50	02:00	02:10	02:20	02:30	RSK
13	Israel	4X6TU	02:00	02:10	02:20	02:30	02:40	U of Tel Aviv
14	Finland	OH2B	02:10	02:20	02:30	02:40	02:50	U oh Helsinki
15	Madeira	CS3B	02:20	02:30	02:40	02:50	00:00	ARRM
16	Argentina	LU4AA	02:30	02:40	02:50	00:00	00:10	RCA
17	Peru	OA4B	02:40	02:50	00:00	00:10	00:20	RCP
18	Venezuela	YV5B	02:50	00:00	00:10	00:20	00:30	RCV

The 10-second, phase-3, message format is: "W6WX dah-dah-dah-dah" — each "dah" lasts a little more than one second. W6WX is transmitted at 100 watts, then each "dah" is attenuated in order, beginning at 100 watts, then 10 watts, then 1 watt, and finally 0.1 watt. *Courtesy CQ Magazine*

Elmer Point: *Which is faster – new-fangled text messaging or old-reliable Morse code? Jay Leno wanted the answer to that question, so on May 13, 2006, Leno invited world text-messaging speed champ Ben Cook of Utah and his friend Jason to appear on* The Tonight Show with Jay Leno *to test their ability against Chip Margelli, K7JA and Ken Miller, K6CTW. Cook told Leno that he'd managed to send a 160-letter message to his friend in 57 seconds. Who won? Chip and Ken, hands down! Margelli sent his message at 29-wpm, but he was once timed sending code at 61.5 words per minute!*

1000101001100100111001000101

G8B09 What do RTTY, Morse code, PSK31 and packet communications have in common?
A. They require the same bandwidth
B. They are digital modes
C. They use on/off keying
D. They use phase shift modulation

RTTY, Morse code, PSK31, and packet are all forms of *digital communication.* Laptop computers with a small analog-to-digital converter, along with the right software, can easily decode these digital communications. Do you have a spare laptop you want to press into service on your new General Class bands? **ANSWER B**

G2E04 Which of the following 20 meter band segments is most often used for most data transmissions?
A. 14.000 - 14.050 MHz
B. 14.070 - 14.100 MHz
C. 14.150 - 14.225 MHz
D. 14.275 - 14.350 MHz

Most *RTTY transmissions* on the *20-meter* band are found between *14.070 to 14.095 MHz*, a 25-kHz RTTY "window." Use the LSB switch to tune in 20-meter RTTY stations using a digital decoder. **ANSWER B**

G2E07 What does the abbreviation "RTTY" stand for?
A. "Returning To You," meaning "your turn to transmit"
B. Radio-Teletype
C. A general call to all digital stations
D. Repeater Transmission Type

Although *radioteletype*, abbreviated *RTTY*, is a relatively outmoded type of communication on worldwide frequencies, there is still some RTTY traffic that you might want to tune in and decode. At night, tune in RTTY on 80 meters near 3600 kHz. During the day, try 15 meters around 21.080 MHz, and 20 meters at 14.080. **ANSWER B**

G2E05 Which of the following describes Baudot RTTY?
A. 7-bit code, with start, stop and parity bits
B. Utilizes error detection and correction
C. 5-bit code, with additional start and stop bits
D. Two major operating modes are SELCAL and LISTEN

Give me FIVE! This is a good way to remember *Baudot* RTTY is a *5 bit code* with an additional start and stop bits. Give me FIVE! **ANSWER C**

Elmer Point: *Down on the digital frequencies you'll hear data activity, and with a simple sound card program you can decode all of the excitement. There are some regular digital contests, too, giving you a great way to begin transmitting and receiving data with other stations needing contact points. Here's a list of some of the bigger data contests:*

- New Years' Day RTTY Contest
- First weekend in January ARRL RTTY Roundup
- First weekend in February Low Power Digital Contest
- Second weekend in February Worldwide RTTY Call Letter Prefix Contest
- Second weekend in March RTTY Sprint
- Third weekend in April PSK-31 Activity weekend
- Third weekend in July North American RTTY Activity weekend
- First weekend in September PSK-31 Contest
- First weekend in October Hellschreiber Contest

G2E06 What is the most common frequency shift for RTTY emissions in the amateur HF bands?
A. 85 Hz   C. 425 Hz
B. 170 Hz   D. 850 Hz

Almost all ham radio RTTY transmissions use the *170-Hz shift*. Commercial broadcast stations on shortwave use larger frequency shifts. **ANSWER B**

G2B06 What minimum frequency separation between 170 Hz shift RTTY signals should be allowed to minimize interference?
A. 60 Hz   C. Approximately 3 kHz
B. 250 to 500 Hz D. 170 Hz

On *RTTY*, similar to CW, stay at least *250 Hz to 500 Hz away* from an ongoing communication. **ANSWER B**

G1C12 What is the maximum authorized bandwidth for RTTY, data or multiplexed emissions using an unspecified digital code transmitted on the 6 and 2 meter bands?
A. 20 kHz
B. 50 kHz
C. The total bandwidth shall not exceed that of a single-sideband phone emission
D. The total bandwidth shall not exceed 10 times that of a CW emission

20 kHz is the authorized bandwidth of a signal within this range on 6- and 2-meters. [97.305(c) and 97.307(f)(5)] **ANSWER A**

G2E01 Which mode should be selected when using a SSB transmitter with an Audio Frequency Shift Keying (AFSK) RTTY signal?

A. USB   C. CW
B. DSB   D. LSB

If you plan to operate radio teleprinter *(RTTY)* using audio-frequency-shift-keying *(AFSK)*, your single sideband transmitter must be switched to *LSB, lower sideband*. **ANSWER D**

You can connect a PSK-31 and RTTY data reader to your radio to decode messages.

G2B10 What should you do to comply with good amateur practice when choosing a frequency for radio-teletype (RTTY) operation?

A. Call CQ in Morse code before attempting to establish a contact in RTTY
B. Select a frequency in the upper end of the phone band
C. Select a frequency in the lower end of the phone band
D. Follow generally accepted band plans for RTTY operation

As a General Class operator, there are many modes of communicating using data. Data communications are found in the upper portion of the *CW and data band plans*, just BELOW frequencies reserved for voice and image communications. **ANSWER D**

G8B11 What part of the 20 meter band is most commonly used for PSK31 operation?

A. At the bottom of the slow-scan TV segment, near 14.230 MHz
B. At the top of the SSB phone segment, near 14.325 MHz
C. In the middle of the CW segment, near 14.100 MHz
D. Below the RTTY segment, near 14.070 MHz

The various digital modes are jockeying for position, especially on the crowded 20-meter band. Unfortunately, collisions occur where PSK31 gets clobbered by a PACTOR II signal, and most of these collisions occur *below the RTTY segment* on 20 meters, *around 14.070 MHz*. Tune down there with your new transceiver and see if you can identify the various data emissions. Listen to my General Class audio theory course to better identify each data signal. **ANSWER D**

G2E09 Where are PSK signals generally found on the 20 meter band?

A. In the low end of the phone band
B. In the high end of the phone band
C. In the weak signal portion of the band
D. Around 14.070 MHz

Plenty of "whistles" are heard *around 14.070 MHz on the 20 meter band*. Listen carefully – do you hear the slight warble? This is the sound of PSK and with relatively inexpensive software you can easily decode this whistle into meaningful text scrolling across your screen. **ANSWER D**

G2B11 What should you do to comply with good amateur practice when choosing a frequency for HF PSK operation?
A. Call CQ in Morse code before attempting to establish a contact in PSK
B. Select a frequency in the upper end of the phone band
C. Select a frequency in the lower end of the phone band
D. Follow generally accepted band plans for PSK operation

At the top portion of the CW and data bands, you will hear the sounds of a whistle, likely a phase shift-keying-transmission *(PSK)*. The *band plan* calls out the individual "slots" where various digital transmissions can take place. **ANSWER D**

G2E02 How many data bits are sent in a single PSK31 character?
A. The number varies   C. 7
B. 5   D. 8

The sounds of *PSK31* take on the characteristics of a steady whistle on the airwaves with just a little warble. It's that variable warble that is part of VARICODE characters represented by a variable-length combination of bits. Just like the name VARICODE implies, *the number of data bits VARIES*. Good reading is the ARRL's *HF Digital Handbook*, edited by Steve Ford, WB8IMY. **ANSWER A**

G2E11 What does the abbreviation "MFSK" stand for?
A. Manual Frequency Shift Keying
B. Multi (or Multiple) Frequency Shift Keying
C. Manual Frequency Sideband Keying
D. Multi (or Multiple) Frequency Sideband Keying

The term *MFSK* stands for *multi frequency shift keying*, and the sound of the 16 tones on the air is quite distinctive. Its best attribute is getting through when the band is fading out to a distant station. **ANSWER B**

G2E10 What is a major advantage of MFSK16 compared to other digital modes?
A. It is much higher speed than RTTY
B. It is much narrower bandwidth than most digital modes
C. It has built-in error correction
D. It offers good performance in weak signal environment without error correction

MFSK 16 will sound much like a calliope on the ham bands. Occupying just over 300 Hz of band width, MFSK 16 is 16 baud with forward error correction. Tuning in an MFSK 16 signal with the correct software requires spot-on accuracy, usually only found on a high frequency transceiver that can read out frequency to the Hz. MFSK 16 is an ideal mode (over RTTY) to overcome the problems of multipath. You will hear *MFSK* on the lower bands like 80 meters because it *will continue to get through when other digital modes fade out*. Look at www.qsl.net/zl1bpu/mfsk **ANSWER D**

G8B08 How is frequency shift related to keying speed in an FSK signal?
A. The frequency shift in hertz must be at least four times the keying speed in WPM
B. The frequency shift must not exceed 15 Hz per WPM of keying speed
C. Greater keying speeds require greater frequency shifts
D. Greater keying speeds require smaller frequency shifts

The *faster* digital and code *emissions* are sent, the *greater* the *bandwidth* these emissions occupy. **ANSWER C**

Mark and Space Frequencies for FSK Spacing

Spacing	Mark	Space
170 Hz	2125 Hz	2295 Hz
170 Hz	1275 Hz	1445 Hz
200 Hz	1270 Hz	1070 Hz Originate
200 Hz	2225 Hz	2025 Hz Answer
425 Hz	2125 Hz	2550 Hz
850 Hz	2975 Hz	2125 Hz
850 Hz	1275 Hz*	2125 Hz

*British Standard, all others U. S. Standard.

Source: *Digital Communications with Packet Radio,* © 1988 Master Publishing, Inc., Niles, IL

G1C08 What is the maximum symbol rate permitted for RTTY emissions transmitted on frequency bands below 28 MHz?
A. 56 kilobaud
B. 19.6 kilobaud
C. 1200 baud
D. 300 baud

Packet emissions below 28 MHz must creep along no faster than *300 baud*. This slow symbol rate is required to minimize bandwidth allocation on the very crowded high-frequency bands. [97.305(c), 97.307(f)(3)] **ANSWER D**

G1C10 What is the maximum symbol rate permitted for RTTY or data emission transmissions on the 10 meter band?
A. 56 kilobaud
B. 19.6 kilobaud
C. 1200 baud
D. 300 baud

The lower the frequency, the lower the baud rate allowed. Don't be disappointed – *1200 baud* is pretty quick on *10 meters*. [97.307(f)(4)] **ANSWER C**

Maximum Symbol (Baud) Rate for Amateur Bands

Amateur Band (meters)	Maximum Symbol Rate (bauds)
160 to 12 m	300 bauds
10 m	1200 bauds
6 and 2 m	19,600 bauds
1.25 and 0.70 m	56,000 bauds
33 cm and higher	Not Specified

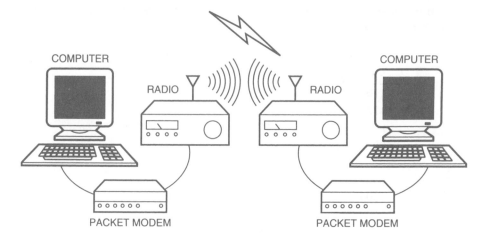

A Packet Radio System

G1C09 What is the maximum symbol rate permitted for packet emission transmissions on the 2 meter band?

A. 300 baud

B. 1200 baud

C. 19.6 kilobaud

D. 56 kilobaud

The higher we go in frequency, the faster we may send our data transmissions. On *2 meters*, we may step up to *19.6 kilobauds*. [97.305(c) and 97.307(f)(5)]

ANSWER C

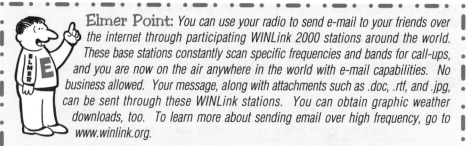

Elmer Point: *You can use your radio to send e-mail to your friends over the internet through participating WINLink 2000 stations around the world. These base stations constantly scan specific frequencies and bands for call-ups, and you are now on the air anywhere in the world with e-mail capabilities. No business allowed. Your message, along with attachments such as .doc, .rtf, and .jpg, can be sent through these WINLink stations. You can obtain graphic weather downloads, too. To learn more about sending email over high frequency, go to www.winlink.org.*

G1C11 What is the maximum symbol rate permitted for RTTY or data emission transmissions on the 6 and 2 meter bands?

A. 56 kilobaud

B. 19.6 kilobaud

C. 1200 baud

D. 300 baud

The next possible answer up from 1200 bauds is *19.6 kilobauds* (kilo means 1000). 19,600 baud is real quick for 6- and 2-meters. [97.305(c) and 97.307(f)(5)]

ANSWER B

G2E03 What part of a data packet contains the routing and handling information?

A. Directory

B. Preamble

C. Header

D. Footer

Within the *header* are *routing addresses* to digipeaters so your packet may be received and relayed over a specific route – even coast-to-coast and worldwide! If you are into APRS (Automatic Position/Packet Reporting System), you can set up your TNC (terminal node controller) header for local or wide-area relays of your position report. **ANSWER C**

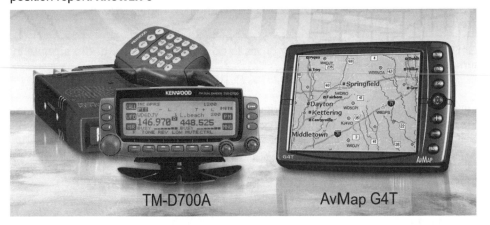

TM-D700A — AvMap G4T

This packet APRS system links the GPS through the ham radio
in Gordo's communications van to send automatic position reports.

G2E08 What segment of the 80 meter band is most commonly used for data transmissions?
A. 3570 - 3600 kHz
B. 3500 - 3525 kHz
C. 3700 - 3750 kHz
D. 3775 - 3825 kHz

When you tune down to the *80 meter* band, especially in the evenings, you will hear the sweet sounds of *data* between *3570 to 3600 kHz*, including the band plan for PACTOR III automatic forwarding stations. When the FCC expanded high frequency phone privileges in December of 2006, digital stations needed to do a little relocation. 3570 to 3600 kHz is the hotbed of data transmissions on the 80 meter band. **ANSWER A**

Amateur packet operations uses your HF radio to link computers on Pactor III

G6C12 What two devices in an amateur radio station might be connected using a USB interface?

A. Computer and transceiver   C. Amplifier and antenna
B. Microphone and transceiver   D. Power supply and amplifier

Two devices at your new General Class station that could be connected using a *USB interface* are your *computer* and that brand new high frequency *transceiver.* **ANSWER A**

G4D10 Which of these connector types is commonly used for audio signals in amateur radio stations?

A. PL-259   C. RCA Phono
B. BNC   D. Type N

Let's look for the correct answer by identifying EACH possible answer. A PL-259 is the coax cable connector that will join your new high frequency ham transceiver. A BNC, along with the SMA, are antenna receptacles on hand-held transceivers. The Type N connector is for UHF and microwave frequencies. In this question, they ask for the *common connector for audio frequencies*, and what you might find on the back of a digital decoder modem is the *RCA phono jack.* **ANSWER C**

G4D08 Which of the following connectors would be a good choice for a serial data port?

A. PL-259   C. Type SMA
B. Type N   D. DB-9

If you have an older laptop and you want to tie it into your new ham radio to decode data, the computer may only offer a *DB-9 serial data port* connector. Most new data decoder programs now ship with a USB connector, no longer the DB-9 data port connector. **ANSWER D**

G2B09 What should you do to comply with good amateur practice when choosing a frequency for Slow-Scan TV (SSTV) operation?

A. Transmit only on lower sideband
B. Transmit your call sign as an SSTV image for 1 minute to ensure a clear frequency
C. Select a frequency in the portion of the band set aside for digital operation
D. Follow generally accepted band plans for SSTV operation

A great way to learn about band plans is to join a ham radio club that may regularly hold technical sessions for new General Class operators to learn their way around their new privileges on the High Frequency bands. Tune in to 14.230 MHz, USB, and listen to the sounds of *slow scan television*. The *band plan* calls this frequency out for ONLY SSTV contacts, and it's easy to decode SSTV by tying your ham rig into your laptop or home computer. SSTV operators welcome reception reports, so give them a call using single sideband and tell them what you are seeing. 14.230 has plenty of activity! **ANSWER D**

Elmer Point: *Want to learn more about how digital electronics work – from DSP systems to your computer, and all those gizmos inside your new HF radio? Get a copy of Basic Digital Electronics by Al Evans. You can pick-up a copy at your local ham radio dealer, on-line at www.w5yi.org, or by calling The W5YI Group at 800-669-9594.*

G2C01 When normal communications systems are not available, what means may an amateur station use to provide essential communications when there is an immediate threat to the safety of human life or the protection of property?
 A. Only transmissions sent on internationally recognized emergency channels
 B. Any means, but only to RACES recognized emergency stations
 C. Any means of radiocommunication at its disposal
 D. Only those means of radiocommunication for which the station is licensed

In an *emergency, any means* and any frequency may be used to get help. You are not confined only to General Class frequencies. In an emergency, you could slide down to the Extra Class portion of the band and signal for help if this is the only area where you hear activity. NEVER would you consider using a hand-held transceiver on a police frequency in an emergency where you might be able to switch to other frequencies to get help. Police channels are absolutely taboo for emergency ham transmissions. **ANSWER C**

In an emergency, authorized hams participating in a RACES
organization may communicate from a police helicopter.

G2C04 When is an amateur station prevented from using any means at its disposal to assist another station in distress?
 A. Only when transmitting in RACES
 B. Only when authorized by the FCC rule
 C. Never
 D. Only on authorized HF frequencies

In an *emergency*, there is *NEVER any restriction* to prevent a station to call out for help. There is NEVER a restriction in an emergency to assist another station who is calling out in distress. [97.405(b)] **ANSWER C**

G2C08 When are you prohibited from helping a station in distress?
A. When that station is not transmitting on amateur frequencies
B. When the station in distress offers no call sign
C. You are never prohibited from helping any station in distress
D. When the station is not another amateur station

In an emergency, you may use your equipment on any frequency to help a station in distress. You are *never prohibited* from helping any station in distress [97.405(b)]
ANSWER C

G2C09 What type of transmissions may an amateur station make during a disaster?
A. Only transmissions when RACES net is activated
B. Transmissions necessary to meet essential communications needs and to facilitate relief actions
C. Only transmissions from an official emergency station
D. Only one-way communications

Resist the temptation to transmit an offer to help, unless the disaster net control operator specifically addresses your station. *Only transmissions directly relating to the relief efforts* are allowed. [97.111(a)(2)] **ANSWER B**

G2C10 Which emission mode must be used to obtain assistance during a disaster?
A. Only SSB   C. Any mode
B. Only SSB and CW   D. Only CW

In an *emergency, any mode* of transmission is allowed. In an emergency, anything goes! **ANSWER C**

Hams are well-known for their work with the Red Cross,
Salvation Army, and others providing emergency communications.

G2C12 What frequency should be used to send a distress call?

A. Whatever frequency has the best chance of communicating the distress message
B. 3873 kHz at night or 7285 kHz during the day
C. Only frequencies that are within your operating privileges
D. Only frequencies used by police, fire or emergency medical services

Before you head out on your adventure, pre-plan what frequency is in use in your expected travel area that could hear your call for help. *Any active frequency* would be a good spot to place a distress call. Do NOT rely on known police, medical, or fire frequencies because most are protected with digital coded inputs that would not be able to decode your carrier. Rather, rely on normal, active, Amateur Radio communication frequencies. **ANSWER A**

G2C11 What information should be given to a station answering a distress transmission?

A. The ITU region and grid square locator of the emergency
B. The location and nature of the emergency
C. The time that the emergency occurred and the local weather
D. The name of the local emergency coordinator

Location is everything. When sending out a distress call, immediately give your location. This way rescue efforts can get started in the correct direction. Maybe your handheld batteries give out, or the boat sinks quickly – first *giving your location*, and then the *nature of the distress*. Ultimately, this will lead to a timely rescue. **ANSWER B**

Hams operating from
an Emergency
Communications trailer

G2C07 What is the first thing you should do if you are communicating with another amateur station and hear a station in distress break in?
A. Continue your communication because you were on frequency first
B. Acknowledge the station in distress and determine what assistance may be needed
C. Change to a different frequency
D. Immediately cease all transmissions

Act quickly to handle a station in distress. Find out WHO is in distress, WHERE they are located, and *WHAT assistance may be needed*. If it's a boat, find out how many persons are on board and instruct everyone to put on their personal floatation device. **ANSWER B**

G1B04 Which of the following must be true before an amateur station may provide news information to the media during a disaster?
A. The information must directly relate to the immediate safety of human life or protection of property and there is no other means of communication available
B. The exchange of such information must be approved by a local emergency preparedness official and transmitted on officially designated frequencies
C. The FCC must have declared a state of emergency
D. Both amateur stations must be RACES stations

Your new General Class ham station is not to be used for routine support of your local television station news. However, *in a disaster*, where a ham on scene tells you the center span of a bridge has just collapsed, it WOULD BE *permissible* for you *to contact your local news agency* and relay this *safety of human life transmission* so motorists don't accidentally fly off the span. [97.113(b)] **ANSWER A**

G2C05 What type of transmission would a control operator be making when transmitting out of the amateur band without station identification during a life threatening emergency?
A. A prohibited transmission
B. An unidentified transmission
C. A third party communication
D. An auxiliary transmission

Even though there is never any restriction to a ham operator using any means to signal in an emergency, that emergency call must always be accompanied by station identification. *Without* a proper *station ID*, the *unidentified transmission* might be mistaken for a hoax call. In an emergency, always ID with as much information as possible. [97.403] **ANSWER B**

G2C02 Who may be the control operator of an amateur station transmitting in RACES to assist relief operations during a disaster?
A. Only a person holding an FCC issued amateur operator license
B. Only a RACES net control operator
C. Only official emergency stations may transmit during a disaster
D. Any control operator when normal communication systems are operational

Radio Amateur Civil Emergency Service (RACES) is a public service by licensed amateur radio operators to provide volunteer communications to government agencies in time of extraordinary need. During periods of RACES activation, licensed amateur radio operators may serve their local government agency. The Federal Emergency Management Agency (FEMA) provides planning guidance and technical assistance for establishing a *RACES* unit at the state and local government level. *Only licensed amateur radio operators* with a current RACES authorization may be the control operator at a RACES station. **ANSWER A**

RACES Logo.

G2C03 When may the FCC restrict normal frequency operations of amateur stations participating in RACES?
　　A. When they declare a temporary state of communication emergency
　　B. When they seize your equipment for use in disaster communications
　　C. Only when all amateur stations are instructed to stop transmitting
　　D. When the President's War Emergency Powers have been invoked
During a time of war, when the President exercises his *War Emergency Powers*, RACES may become the ONLY communications allowed via amateur radio. All other non-RACES ham operators would be ordered to stop transmitting – only RACES operators could remain on the air. [97.407(b)] **ANSWER D**

Website Resources

▼ IF YOU'RE LOOKING FOR	▼ THEN VISIT
All About RACES	www.RACES.net
News on Emergency Groups	www.N4KSS.net/Reflectors.html
All about ARES	www.QSL.net/ARES
Military Radio Groups	www.NAVYMARS.org
More About Joining ARES	www.ARRL.org/ARES
ARRL Emergency Volunteers	www.ARRL.org/Volunteer

The Effect of the Ionosphere on Radio Waves

To help you with the questions on radio wave propagation, here is a brief explanation on the effect the ionosphere has on radio waves.

The ionosphere is the electrified atmosphere from 40 miles to 400 miles above the Earth. You can sometimes see it as "northern lights." It is charged-up daily by the Sun, and does some miraculous things to radio waves that strike it. Some radio waves are absorbed during daylight hours by the ionosphere's D layer. Others are bounced back to Earth. Yet others penetrate the ionosphere and never come back again. The wavelength of the radio waves determines whether the waves will be absorbed, refracted, or will penetrate. Here's a quick way to memorize what the different layers do during day and nighttime hours:

The D layer is about 40 miles up. The D layer is a Daylight layer; it almost disappears at night. D for Daylight. The D layer absorbs radio waves between 1 MHz to 7 MHz. These are long wavelengths. All others pass through.

The E layer is also a daylight layer, and it is very Eccentric. E for Eccentric. Patches of E layer ionization may cause some surprising reflections of signals on both high frequency as well as very-high frequency. The E layer height is usually 70 miles.

The F1 layer is one of the layers farthest away. The F layer gives us those Far away signals. F for Far away. The F1 layer is present during daylight hours, and is up around 150 miles. The F2 layer is also present during daylight hours, and it gives us the Furthest range. The F2 layer is 250 miles high, and it's the best for the Farthest range on medium and short waves. The F2 layer is strongest in the summer months. During winter months, both the F1 and F2 layers may become unpredictable, but always strong enough to support exciting skywaves! At nighttime, the F1 and F2 layers combine to become just the F layer at 180 miles. This F layer at nighttime will usually bend radio waves between 1 MHz and 15 MHz back to earth. At night, the D and E layers disappear.

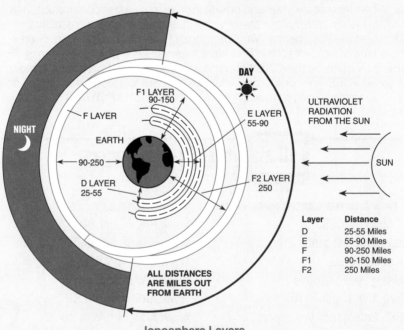

Ionosphere Layers

Source: *Antennas — Selection and Installation*, © 1986, Master Publishing, Inc., Niles, Illinois

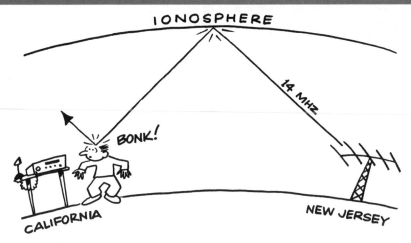

Skywave Excitement

G3C03 Why is the F2 region mainly responsible for the longest distance radio wave propagation?

A. Because it is the densest ionospheric layer
B. Because it does not absorb radio waves as much as other ionospheric regions
C. Because it is the highest ionospheric region
D. All of these choices are correct

The *higher* the *altitude* of the ionospheric region refracting the high-frequency radio waves, the greater the radio range. **ANSWER C**

G3B09 What is the maximum distance along the Earth's surface that is normally covered in one hop using the F2 region?

A. 180 miles
B. 1,200 miles
C. 2,500 miles
D. 12,000 miles

The *F2 layer* is our highest reflective ionospheric region, approximately 250 miles up. This gives worldwide signals their furthest bounce, generally about *2500 miles*. **ANSWER C**

G3C02 When can the F2 region be expected to reach its maximum height at your location?

A. At noon during the summer
B. At midnight during the summer
C. At dusk in the spring and fall
D. At noon during the winter

All of the ionospheric regions are influenced by ultraviolet radiation from the Sun. The *F2 region* is at its *ultimate height at noon during the summer*. **ANSWER A**

G3C04 What does the term "critical angle" mean as used in radio wave propagation?

A. The long path azimuth of a distant station
B. The short path azimuth of a distant station
C. The lowest takeoff angle that will return a radio wave to the Earth under specific ionospheric conditions
D. The highest takeoff angle that will return a radio wave to the Earth under specific ionospheric conditions

---

Have you ever skipped stones on a lake? There is an angle that you cannot exceed where the stone doesn't skip, but rather penetrates into the water. Radio waves in the ionosphere act similarly; there is a point – the highest take-off angle – that cannot be exceeded or a radio wave will not refract back to Earth. Just remember *"highest take-off angle."* **ANSWER D**

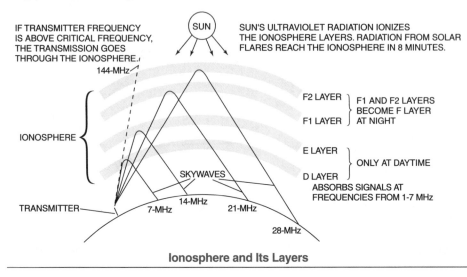

IF TRANSMITTER FREQUENCY IS ABOVE CRITICAL FREQUENCY, THE TRANSMISSION GOES THROUGH THE IONOSPHERE.

144-MHz

SUN

SUN'S ULTRAVIOLET RADIATION IONIZES THE IONOSPHERE LAYERS. RADIATION FROM SOLAR FLARES REACH THE IONOSPHERE IN 8 MINUTES.

F2 LAYER
F1 LAYER
} F1 AND F2 LAYERS BECOME F LAYER AT NIGHT

IONOSPHERE

E LAYER
D LAYER
} ONLY AT DAYTIME

SKYWAVES

ABSORBS SIGNALS AT FREQUENCIES FROM 1-7 MHz

TRANSMITTER
7-MHz
14-MHz
21-MHz
28-MHz

Ionosphere and Its Layers

G3B08 What does MUF stand for?
    A. The Minimum Usable Frequency for communications between two points
    B. The Maximum Usable Frequency for communications between two points
    C. The Minimum Usable Frequency during a 24 hour period
    D. The Maximum Usable Frequency during a 24 hour period

The *maximum usable frequency (MUF)* peaks during our day in the morning hours to the east, for working Europe, and in the afternoon and evening hours, to the west, working Asia. This is the fun of working ham radio skywaves — the ionosphere and the maximum usable frequency will constantly give us some excitement if we just keep tuned in. **ANSWER B**

G3B05 What usually happens to radio waves with frequencies below the maximum usable frequency (MUF) when they are sent into the ionosphere?
    A. They are bent back to the Earth
    B. They pass through the ionosphere
    C. They are completely absorbed by the ionosphere
    D. They are bent and trapped in the ionosphere to circle the Earth

Frequencies below the maximum usable frequency are *bent back to Earth* by the ionosphere. For maximum range, operate as close to MUF as possible. **ANSWER A**

G3B03 Which of the following guidelines should be selected for lowest attenuation when transmitting on HF?
    A. Select a frequency just below the MUF
    B. Select a frequency just above the LUF
    C. Select a frequency just below the critical frequency
    D. Select a frequency just above the critical frequency

The term "lowest attenuation" means least amount of signal fading. The trick is to try to operate on the highest high frequency ham band that gives you skywave propagation to somewhere else in the country or the world. Operating *just below the maximum usable frequency* will lead to some extraordinary crystal clear communications. **ANSWER A**

G3B04 What is a reliable way to determine if the maximum usable frequency (MUF) is high enough to support 28-MHz propagation between your station and Western Europe?
    A. Listen for signals on a 28 MHz international beacon
    B. Send a series of dots on the 28 MHz band and listen for echoes from your signal
    C. Check the strength of TV signals from Western Europe
    D. Listen to WWV propagation signals on the 28 MHz band

Tune between 28.2- to 28.3-MHz on the 10-meter band, *listen for continuous beacons*, and copy down the call signs. If you hear stations *between 28.2 to 28.3-MHz*, the band is wide open and propagation is taking place. **ANSWER A**

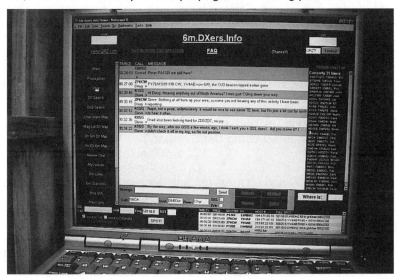

There are websites that provide skywave DX conditions.

G3B01 Which band should offer the best chance for a successful contact if the maximum usable frequency (MUF) between the two stations is 22 MHz?

A. 10 meters	C. 20 meters
B. 15 meters	D. 40 meters

The next ham band down from 22 MHz MUF is the *21 MHz, 15 meter ham band*. This should give us some successful skywave contacts. **ANSWER B**

G3B02 Which band should offer the best chance for a successful contact if the maximum usable frequency (MUF) between the two stations is 16 MHz?

A. 80 meters	C. 20 meters
B. 40 meters	D. 2 meters

The ham band down from 16 MHz MUF is the *14 MHz, 20 meter band*. 20 meters should give you successful skywave contacts. **ANSWER C**

G3B12 What factors affect the maximum usable frequency (MUF)?
A. Path distance and location
B. Time of day and season
C. Solar radiation and ionospheric disturbance
D. All of these choices are correct

Getting a signal halfway around the Earth, operating at just below the maximum usable frequency, is a ham radio tradition. But there are plenty of things to consider – solar activity, day or night, fall or summer, and how far away the other station is located. *All of these factors affect the MUF.* **ANSWER D**

G3C11 Which of the following is true about ionospheric absorption near the maximum usable frequency (MUF)?
A. Absorption will be minimum
B. Absorption is greater for vertically polarized waves
C. Absorption approaches maximum
D. Absorption is greater for horizontally polarized waves

Ham operators take great pride in being able to choose a band right at the *maximum usable frequency*. This will allow them a powerful skywave skip signal with *minimum absorption*. So how do you find out where the MUF is? Dial into various high-frequency bands to see which is the highest band where there is skywave activity pouring in. If you hear plenty of skywave signals on 15 meters, but none on 10 and 12 meters, 15 meters is probably your best choice. But hey, a quick call on 12 meters or 10 meters might give you some surprising results, too! **ANSWER A**

G3B10 What is the maximum distance along the Earth's surface that is normally covered in one hop using the E region?
A. 180 miles   C. 2,500 miles
B. 1,200 miles   D. 12,000 miles

The E layer is between 50 and 90 miles up, and because it's closer to Earth, high-frequency waves don't bounce as far as they do off of the F layer. *E skip* is approximately *1200 miles* and, during the summertime, "sporadic E" may sometimes "short skip" in as close as 600 miles. **ANSWER B**

G3B14 Which of the following is a good indicator of the possibility of sky-wave propagation on the 6 meter band?
A. Short hop sky-wave propagation on the 10 meter band
B. Long hop sky-wave propagation on the 10 meter band
C. Severe attenuation of signals on the 10 meter band
D. Long delayed echoes on the 10 meter band

Now that you are upgrading from Technician to General, don't abandon all the excitement on 6 meters. In fact, your new General Class privileges on 10 meters will help you forecast when we might have an upcoming 6-meter band opening. When you begin to make *contact via skywaves on 10 meters* with stations less than 300 miles away, this *indicates* an extremely strong Sporadic E skip "cloud" out there in the ionosphere, and *E skip* on 6 meters and maybe even 2 meters will more than likely be possible within the next half hour! **ANSWER A**

G3C01 Which of the following ionospheric layers is closest to the surface of the Earth?

A. The D layer
B. The E layer
C. The F1 layer
D. The F2 layer

The Darn D layer! It is the layer closest to the surface of the Earth and during daylight hours it is usually responsible for absorbing ham radio medium-frequency skywave signals. **ANSWER A**

Altitudes in Miles of Ionospheric Layers

Layers	Day Summer	Winter	Night
F2	>250		
F1	90-150		
F		90-150	90-250
E	55-90	55-90	
D	40	40	

G3C12 Which ionospheric layer is the most absorbent of long skip signals during daylight hours on frequencies below 10 MHz?

A. The F2 layer
B. The F1 layer
C. The E layer
D. The D layer

The 40-meter band is a great one for 500-mile, daylight skywave contacts. Even though they may fade in and out a little bit, they are almost always there from Sun-up to around 4:00 p.m. local time. After 4:00 p.m., the D-layer associated with absorption begins to disappear, and the 40-meter band begins to go "long." Your 500-mile buddies will disappear, and the next thing you hear are stations a couple thousand miles away pouring in. Then, as the 40-meter band continues on well into the night, in come the dreaded megawatt, foreign, double-sideband, broadcast stations that share our frequencies, too. What you will hear at night and in the early morning hours on 40 meters are extremely loud whistles from foreign broadcast, double-sideband, full-carrier stations, and the art of operating 40 meters in the early morning hours is finding a spot to dodge the foreign broadcast heterodynes. But during the day, *the Darn D Layer* causes fades in local 500-mile contacts. **ANSWER D**

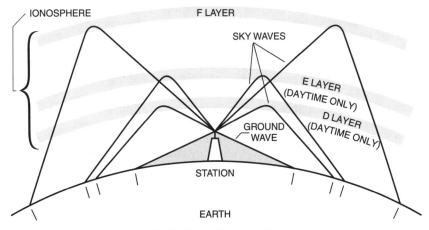

Radio Wave Propagation

Source: *Mobile 2-Way Radio Communications*, G. West, © 1993, Master Publishing, Inc.

G3C05 Why is long distance communication on the 40, 60, 80 and 160 meter bands more difficult during the day?
A. The F layer absorbs these frequencies during daylight hours
B. The F layer is unstable during daylight hours
C. The D layer absorbs these frequencies during daylight hours
D. The E layer is unstable during daylight hours

The "Darn D" layer does more harm than good to medium-frequency and high-frequency signals. During daylight hours, SSB and CW operation on 160- and 80-meters is confined to ground wave coverage. On 40 meters, daytime skip distances are generally no greater than 600 miles when the *D layer* is doing its thing *absorbing MF and HF signals.* **ANSWER C**

G3B07 What does LUF stand for?
A. The Lowest Usable Frequency for communications between two points
B. The Longest Universal Function for communications between two points
C. The Lowest Usable Frequency during a 24 hour period
D. The Longest Universal Function during a 24 hour period

The term "lowest usable frequency" refers specifically to skywave stations attempting contact. On high frequency ground waves, out to 20 miles, you will get through no matter what the ionosphere is doing. But if you and a buddy want to stay in touch on the 75 meter ham band, separated by 500 miles, you may find early morning contacts loud and clear, but at high noon, the *lowest usable frequency* is many megahertz above where you plan to operate, and no contact will be made. But hang around – the *LUF* will begin to drop quickly around sundown. **ANSWER A**

G3B06 What usually happens to radio waves with frequencies below the lowest usable frequency (LUF)?
A. They are bent back to the Earth
B. They pass through the ionosphere
C. They are completely absorbed by the ionosphere
D. They are bent and trapped in the ionosphere to circle the Earth

During daylight hours, the lowest usable frequency for a skywave contact may be around 3 MHz and anything below that is *absorbed.* This means the 160 meter band, just under 2 MHz, will not be usable for daylight skywave contacts. But as soon as the sun goes down, hang on for some great DX contacts! **ANSWER C**

G3B11 What happens to HF propagation when the lowest usable frequency (LUF) exceeds the maximum usable frequency (MUF)?
A. No HF radio frequency will support communications over the path
B. HF communications over the path are enhanced at the frequency where the LUF and MUF are the same
C. Double hop propagation along the path is more common
D. Propagation over the path on all HF frequencies is enhanced

When the lowest usable frequency *(LUF)* jumps up and actually *exceeds* the *maximum usable frequency*, high-frequency *radio communications* along a specific ionospheric signal path will *disappear.* This sometimes occurs with increased geomagnetic activity from Sunspots. **ANSWER A**

G3C09 What type of radio wave propagation allows a signal to be detected at a distance too far for ground wave propagation but too near for normal sky wave propagation?

A. Ground wave
B. Scatter

C. Sporadic-E skip
D. Short-path skip

Backscatter communications is one way to reach a station that is in that zone of no-reception – the skip zone. When I communicate from southern California to Seattle, San Francisco is in my skip zone and will not receive my signals. But if I aim my beam antenna west toward Hawaii, some of my signal is backscattered into the Bay Area, giving me communications to a station that is too far for ground wave, and too close for normal sky waves. **ANSWER B**

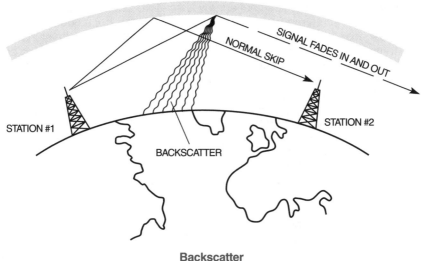

NORMAL SKIP
SIGNAL FADES IN AND OUT
STATION #1
STATION #2
BACKSCATTER

Backscatter

G3C06 What is a characteristic of HF scatter signals?

A. They have high intelligibility
B. They have a wavering sound
C. They have very large swings in signal strength
D. All of these choices are correct

High frequency scatter communications bounce a portion of your signal off of densely ionized patches in the ionosphere. Since the ionosphere is constantly in motion, the signals will fade in and out, much like ocean waves, resulting in a *wavering sound*. **ANSWER B**

G3C07 What makes HF scatter signals often sound distorted?

A. The ionospheric layer involved is unstable
B. Ground waves are absorbing much of the signal
C. The E-region is not present
D. Energy is scattered into the skip zone through several radio wave paths

The wavy sound of HF scatter signals, especially backscatter, is caused by the *signal* being *reflected* back *through several radio-wave paths*, creating multi-path distortion. **ANSWER D**

G3C10 Which of the following might be an indication that signals heard on the HF bands are being received via scatter propagation?

    A. The communication is during a sunspot maximum

    B. The communication is during a sudden ionospheric disturbance

    C. The signal is heard on a frequency below the maximum usable frequency

    D. The signal is heard on a frequency above the maximum usable frequency

If you chose a *frequency slightly above the maximum usable frequency*, you can sometimes take advantage of the ionosphere to scatter your communications to an area that normally would not hear radiowave reflection from the ionosphere. **ANSWER D**

G3C08 Why are HF scatter signals in the skip zone usually weak?

    A. Only a small part of the signal energy is scattered into the skip zone

    B. Signals are scattered from the troposphere, which is not a good reflector

    C. Propagation is through ground waves, which absorb most of the signal energy

    D. Propagation is through ducts in the F region, which absorb most of the energy

During periods of scatter communications, *only a fraction of the original signal is scattered back* to those stations too far for ground-wave reception, yet too close for the main part of your signal being reflected by the ionosphere. Some of the scattered signals are refracted back into your skip zone, so all you get is a very weak incoming skywave. **ANSWER A**

G2D04 What is an azimuthal projection map?

    A. A world map projection centered on the North Pole

    B. A world map projection centered on a particular location

    C. A world map that shows the angle at which an amateur satellite crosses the equator

    D. A world map that shows the number of degrees longitude that an amateur satellite appears to move westward at the equator with each orbit

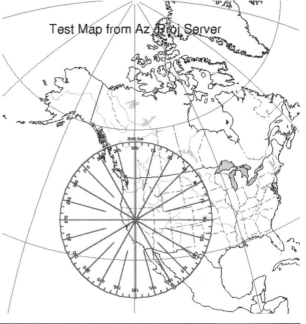

Long-range communications do not necessarily go in straight lines. When we navigate our signals around the world, we need *a chart that takes into account our location* and the curvature of the earth. The azimuthal map will determine the shortest path between your station and that rare DX station. If you want to see a sample of an azimuthal projection map, visit www.wm7d.net/az_proj. **ANSWER B**

Ham operators with a beam antenna will use an azimuthal map like this one to determine short path and long path headings to reach DX stations.

G3B13 How might a sky-wave signal sound if it arrives at your receiver by both short path and long path propagation?
- A. Periodic fading approximately every 10 seconds
- B. Signal strength increased by 3 dB
- C. The signal will be cancelled causing severe attenuation
- D. A well-defined echo can be heard

Do you have my audio course? If so, you can hear the distinctive sound of simultaneous short-path and *long-path* reception. Incoming *signals will have a well-defined echo* because the short-path signal is coming to you from a much shorter distance than the long-path signal coming all the way around the globe. **ANSWER D**

G2D06 How is a directional antenna pointed when making a "long-path" contact with another station?
- A. Toward the rising sun
- B. Along the Gray Line
- C. 180 degrees from its short-path heading
- D. Toward the North

Some ionospheric conditions may allow you to establish communications with a distant station on General Class worldwide frequencies over a longer path around the world than the direct short path. If you hear the station with an echo, try turning your beam antenna *180 degrees in the opposite direction from the short path* direction to see whether or not the station will come in better on the long path. The echo you hear is the difference in the transmission delay between long path and short path transmissions. **ANSWER C**

G3A11 How long is the typical sunspot cycle?
- A. Approximately 8 minutes
- B. Between 20 and 40 hours
- C. Approximately 28 days
- D. Approximately 11 years

Every *11 years*, the sun goes through a *complete solar cycle* variation in sunspot numbers. For the next couple of years, we will go from few sunspots to MORE sunspot activity. We are just beginning to climb up solar cycle 24. And even though we may be at the bottom of solar cycle 23, or the beginning bottom of solar cycle 24, there will be plenty of worldwide high frequency activity throughout the HF bands. **ANSWER D**

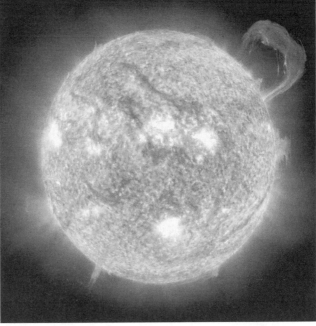

Solar flares and sunspots affect radiowave propagation

Photo courtesy of N.A.S.A.

G3A10 What is the sunspot number?

A. A measure of solar activity based on counting sunspots and sunspot groups
B. A 3 digit identifier which is used to track individual sunspots
C. A measure of the radio flux from the sun measured at 10.7 cm
D. A measure of the sunspot count based on radio flux measurements

We have been studying the surface of the Sun since 1610. The Zurich Observatory was built in 1749, and 100 years later began to regularly record Sunspot numbers. Every 11 years the Sun goes through a period of huge solar eruptions of charged particles. These charged electrons and ions from the Sun become a solar wind that streams into our upper magnetosphere, which sometimes enhances ham radio transmissions on VHF bands, and sometimes soaks-up normal skywave contacts on high frequency. A report of the *sunspot number* refers to a *daily index of sunspot activity*. **ANSWER A**

G3A09 What is the effect on radio communications when sunspot numbers are high?

A. High-frequency radio signals become weak and distorted
B. Frequencies above 300 MHz become usable for long-distance communication
C. Long-distance communication in the upper HF and lower VHF range is enhanced
D. Long-distance communication in the upper HF and lower VHF range is diminished

We are just beginning an upward climb of a new solar cycle, cycle 24. We can expect to see the cycle peak around 2012, and plan for *enhanced*, fun *operation* throughout the entire solar cycle climb on high frequencies. Sunspots will become more common as we approach this solar cycle peak in a few more years.
ANSWER C

G3A18 If the HF radio-wave propagation (skip) is generally good on the 24-MHz and 28-MHz bands for several days, when might you expect a similar condition to occur?

A. 7 days later
B. 14 days later
C. 28 days later
D. 90 days later

You will find *recurring sky-wave conditions about 28 days later*, as the Sun makes a complete rotation. Ham radio propagation forecasts are generally quite accurate. Subscribe to Amateur Radio monthly publications, which always include propagation forecasts. **ANSWER C**

G3A04 What is measured by the solar flux index?

A. The density of the Sun's magnetic field
B. The radio energy emitted by the Sun
C. The number of sunspots on the side of the Sun facing the Earth
D. A measure of the tilt of the Earth's ionosphere on the side toward the Sun

We can actually measure the *radio noise energy emitted by the Sun* with special receiving equipment. You can also test your VHF and UHF antenna system by aiming these antennas at the Sun and listening to Sun noise. **ANSWER B**

G3A05 What is the solar-flux index?

A. A measure of the highest frequency that is useful for ionospheric propagation between two points on the Earth

B. A count of sunspots which is adjusted for solar emissions

C. Another name for the American sunspot number

D. A measure of solar activity at 10.7 cm

The *solar-flux index is measured daily* in Ottawa, Canada. Measurements are made on the frequency of 2800 MHz late every afternoon. You may tune into the radio propagational solar activity reports transmitted by WWV at 18 minutes past the hour. Frequencies of 10- and 15-MHz will give you best reception during the day, and 5-MHz may give you best reception at night. **ANSWER D**

G3A12 What is the K-index?

A. An index of the relative position of sunspots on the surface of the Sun

B. A measure of the short term stability of the Earth's magnetic field

C. A measure of the stability of the Sun's magnetic field

D. An index of solar radio flux measured at Boulder, Colorado

Earth is surrounded by the magnetosphere that acts as a barrier from some of the charged particles coming from eruptions on the Sun. The magnetosphere occasionally develops holes and cracks that may allow tremendous amounts of solar energy to disturb our natural geomagnetic stability. Some of this energy that penetrates the magnetosphere can be seen as auroras. A low *K index* means good, *stable* high-frequency propagation. **ANSWER B**

Elmer Point: *Your new HF General Class radio also offers full shortwave reception. Try 10- or 15-MHz to hear the WWV time ticks. At 18 minutes past the hour, listen to your latest ionospheric and solar weather reports! A K index between 1 to 4, and an A index of 0 to 7 means the ionosphere is quiet and very predictable for long-range skywaves. But a K index of 5 or 6 with an A index of 30 to 49 indicates that HF band conditions will become unpredictable and unstable. And hang on to your receivers – a K index of 7 to 9 or an A index of 50 and higher means the HF bands will likely be unusable for regular skywave contacts due to a major solar storm with strong major geomagnetic activity disrupting HF signals. With high K and A indexes, time to go outside after dark and look for an aurora! Here's a summary of K and A Index readings:*

K Index	A Index	HF Skip Conditions
K1 – K4	A0 – A7	Bands are normal
K4	A8 – A15	Bands are unsettled
K4	A16 – A30	Bands are unpredictable
K5	A30 – A50	Lower bands are unstable
K6	A50 – A99	Few skywaves below 15 MHz
K7 – K9	A100 – A400	Radio blackout is likely.
		Go fishing or watch for an aurora.

G3A13 What is the A-index?
A. An index of the relative position of sunspots on the surface of the sun
B. The amount of polarization of the Sun's electric field
C. An indicator of the long term stability of the Earth's geomagnetic field
D. An index of solar radio flux measured at Boulder, Colorado

The A-Index is a 24 hour averaging of the planetary *K-Index*, and is a great indicator of *LONG TERM stability of the Earth's geomagnetic field*. **ANSWER C**

G3A03 How long does it take the increased ultraviolet and X-ray radiation from solar flares to affect radio-wave propagation on the Earth?
A. 28 days
B. Several hours depending on the position of the Earth in its orbit
C. Approximately 8 minutes
D. 20 to 40 hours after the radiation reaches the Earth

Ultraviolet radiation travels at the speed of light. It takes about *8 minutes for* sunlight and *ultraviolet rays to reach the Earth's* ionosphere. Sunspots may quickly appear, and heavy Sunspot activity may affect worldwide propagation for up to 3 days. **ANSWER C**

G3A06 What is a geomagnetic disturbance?
A. A sudden drop in the solar-flux index
B. A shifting of the Earth's magnetic pole
C. Ripples in the ionosphere
D. A significant change in the Earth's magnetic field over a short period of time

When there are *major flare-ups* on the Sun, it *will change the Earth's magnetic field* over a short period of time, affecting worldwide radio waves. There is little or no effect on VHF or UHF signals. **ANSWER D**

G3A02 What effect does a Sudden Ionospheric Disturbance (SID) have on the daytime ionospheric propagation of HF radio waves?
A. It disrupts higher-latitude paths more than lower-latitude paths
B. It disrupts signals on lower frequencies more than those on higher frequencies
C. It disrupts communications via satellite more than direct communications
D. None, because only areas on the night side of the Earth are affected

The *lower bands*, such as 160-, 80-, 40-, and even 20-meters, *become so noisy* that it is impossible to hear any distant signals coming in from sky waves. **ANSWER B**

G3A07 Which latitudes have propagation paths that are more sensitive to geomagnetic disturbances?
A. Those greater than 45 degrees North or South latitude
B. Those between 5 and 45 degrees North or South latitude
C. Those at or very near to the equator
D. All paths are affected equally

Do you live *north of 45 degrees north*? If you do, your location will be more sensitive to geomagnetic disturbances than for the rest of us down here at latitude 33 degrees. **ANSWER A**

G3A08 What can be an effect of a geomagnetic storm on radio-wave propagation?
   A. Improved high-latitude HF propagation
   B. Degraded high-latitude HF propagation
   C. Improved ground-wave propagation
   D. Improved chances of UHF ducting

During periods of major geomagnetic disturbances, *high-frequency propagation over high-latitude paths will be degraded*. 6 meters VHF might be hopping!
ANSWER B

G3A16 What is a possible benefit to radio communications resulting from periods of high geomagnetic activity?
   A. Aurora that can reflect VHF signals
   B. Higher signal strength for HF signals passing through the polar regions
   C. Improved HF long path propagation
   D. Reduced long delayed echoes

During periods of high geomagnetic activity, beautiful auroras may be viewed at night, away from city lights, as far south as latitude 37. The aurora itself is composed of gases in our upper atmosphere getting bombarded by coronal mass ejections, kind of like a neon light. Depending on which gas gets stirred up determines what color Northern Lights you may see. A trip to Alaska during an equinox will usually lead to some awe-inspiring Northern Lights viewing. You will definitely be impressed by an aurora. The *aurora can* also *reflect VHF signals*, too, adding to the visual excitement. **ANSWER A**

Geomagnetic disturbances caused by the Sun result in the Northern Lights.

G3A01 What can be done at an amateur station to continue communications during a sudden ionospheric disturbance?

A. Try a higher frequency   C. Try a different antenna polarization

B. Try the other sideband   D. Try a different frequency shift

During a SID (sudden ionospheric disturbance), bands like 40- and 20-meters develop extremely-high noise levels, sometimes greater than S9. *Switch up to 15- or 10-meters* to see if you can hear distant stations during this "radio blackout" period. **ANSWER A**

G3A15 How long does it take charged particles from Coronal Mass Ejections to affect radiowave propagation on the Earth?

A. 28 days   C. The effect is instantaneous

B. 14 days   D. 20 to 40 hours

We can spot a sunspot in the amount of time it takes light to travel from the Sun to the Earth – 8 minutes. But this specific question asks about the slower moving *sunspot charged particles* that lumber toward Earth as part of the solar wind, taking as much as *20 to 40 hours* to begin to disturb radiowave propagation down here on the ham bands. This means we have an almost 2 days "heads up" alert that band conditions may be changing. Tune into WWV at 18 minutes past the hour and listen to the 30-second solar report.

WWV 5 MHz	Best at night
WWV 10 MHz	Day and night
WWV 15 MHz	Best days
WWV & WWVH 20 MHz	Some days

You can usually tune in WWV 10 MHz during the days, but only before a big sunspot event finally hits here on Earth 20 to 40 hours later – when disruption occurs, WWV sometimes will fade out completely! **ANSWER D**

G3A14 How are radio communications usually affected by the charged particles that reach the Earth from solar coronal holes?

A. HF communications are improved

B. HF communications are disturbed

C. VHF/UHF ducting is improved

D. VHF/UHF ducting is disturbed

A solar coronal hole can be seen as a dark spot on the face of the Sun emitting charged particles into the solar wind. If you took your thermometer up there, there would be an extreme temperature drop within the eruption. You can actually track sunspots as they rotate around the Sun on a 27.5-day cycle. There are usually a pair of sunspots – a pair with a positive magnetic north field, and a pair with a negative south field. Just ask Galileo – he was the first to observe these sunspots in 1610. *Sunspots* will normally *disrupt high-frequency communications*, but for the joy of 6-meter and 2-meter operators, sunspots can create aberrations in the magnetosphere that will cause VHF long-range auroral band openings. But on HF, coronal holes may lead to poor band conditions. **ANSWER B**

G3A19 Which frequencies are least reliable for long distance communications during periods of low solar activity?

A. Frequencies below 3.5 MHz   C. Frequencies at or above 10 MH
B. Frequencies near 3.5 MHz   D. Frequencies above 20 MHz

Now that we are at the bottom of solar cycle 23 and just beginning to enter solar cycle 24, don't expect much regular long-distance communications on *frequencies above 20 MHz.* **ANSWER D**

G3A17 At what point in the solar cycle does the 20 meter band usually support worldwide propagation during daylight hours?

A. At the summer solstice
B. Only at the maximum point of the solar cycle
C. Only at the minimum point of the solar cycle
D. At any point in the solar cycle

When you earn your new General Class license, you won't need to worry that we are at the bottom of the solar cycle and are a few years away from the peak of cycle 24. Your favorite band will probably be *20 meters*, because it *propagates skywaves at any point in the solar cycle.* **ANSWER D**

Website Resources

▼ IF YOU'RE LOOKING FOR	▼ THEN VISIT
icom radio site for dx comms	www.dxer.com
World Radio magazine	www.wr6wr.com
antennas, DSP, and more	www.amcominc.com
THE place for QSO cards	www.w4mpy.com
DX reference guide	www.ac6v.com
amateur radio satellite operations	www.amsat.org
the latest news about ham radio	www.arnewsline.org
microphones, headsets, and more	www.heilsound.com
digital ham radio accessories and more	www.packetradio.com
Los Angeles disaster comms service	www.lacdcs.org
Disaster preparedness & emergency comms info	www.fema.gov
NASA research site with cool articles	www.grc.nasa.gov
DSP, antenna analyzers, and more	www.timewave.com
HF modems for digital modes	www.halcomm.com
HF modems and gear for digital modes	www.kantronics.com
digital amateur radio organization	www.tapr.org
azimuthal maps, solar information, and more	www.wm7d.net

Your HF Transmitter

G8A01 What is the name of the process that changes the envelope of an RF wave to convey information?

A. Phase modulation
B. Frequency modulation
C. Spread Spectrum modulation
D. Amplitude modulation

A type of modulation that will change the amplitude (envelope) of an RF wave is called *amplitude modulation (AM).* **ANSWER D**

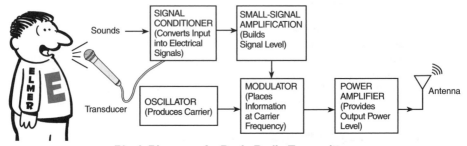

Block Diagram of a Basic Radio Transmitter
Source: *Basic Communications Electronics,* Hudson & Luecke,
© 1999, Master Publishing, Inc., Niles, Illinois

G8A05 What type of transmission varies the instantaneous power level of the RF signal to convey information?

A. Frequency shift keying
B. Pulse modulation
C. Frequency modulation
D. Amplitude modulation

The instantaneous amplitude (power level) of the signal varies with *amplitude modulation.* **ANSWER D**

G7A07 Which circuit is used to combine signals from the carrier oscillator and speech amplifier and send the result to the filter in a typical single-sideband phone transmitter?

A. Mixer
B. Detector
C. IF amplifier
D. Balanced modulator

In an SSB transceiver, the *balanced modulator* processes the signal from the carrier oscillator and the speech amplifier and sends it on to the filter. **ANSWER D**

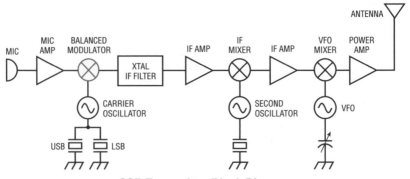

SSB Transmitter Block Diagram

Elmer Point: *Ham radio single sideband transceivers in the $1800 range will likely offer adjustable digital signal processing filters to perfectly set transmit audio characteristics and bandwidth as well as receive bandwidth selections. This is why I always recommend buying new HF transceivers that now include new digital network filters.*

G8A12 What signal(s) would be found at the output of a properly adjusted balanced modulator?
   A. Both upper and lower sidebands
   B. Either upper or lower sideband, but not both
   C. Both upper and lower sidebands and the carrier
   D. The modulating signal and the unmodulated carrier

Just as the name implies, the "balanced modulator" in your new single-sideband General Class ham transceiver includes *BOTH the upper and lower sidebands* at its output with the carrier reduced to near zero. Then the signal goes into a sideband filter, and you end up with normally lower sideband on 160-, 75-, and 40-meters, with upper sideband found on the brand new 60-meter, 5 MHz channels, and upper sideband naturally on 20-, 17-, 15-, 12-, 10-, and 6-meters, too. Only brand new equipment offers 60-meter, 5 MHz capabilities, so check with your local radio store to see how you can modify an older SSB to work on any one of the new 5 channels at 5 MHz, double checking that your radio is switched to upper sideband. **ANSWER A**

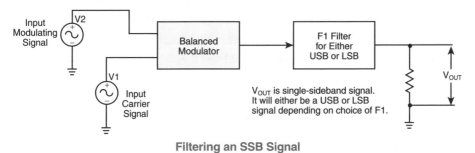

Filtering an SSB Signal

G7A06 Which of the following might be used to process signals from the balanced modulator and send them to the mixer in a single-sideband phone transmitter?
   A. Carrier oscillator   C. IF amplifier
   B. Filter   D. RF amplifier

The balanced modulator will produce an upper and lower sideband signal, with the carrier balanced out into suppression. The upper and lower sideband signal is then fed into a narrow *filter* network which cancels out either the upper or the lower sideband. Modern high frequency ham transceivers now incorporate menu items that allow you to select the shape of the remaining sideband signal, which is then fed to the driver amplifier. The more expensive the ham radio base station, the more elaborate the filter networks. **ANSWER B**

G8A07 Which of the following phone emissions uses the narrowest frequency bandwidth?

A. Single sideband
B. Double sideband
C. Phase modulation
D. Frequency modulation

The modern high frequency ham radio transceiver may include *single sideband* filter networks which decrease the amount of spectrum that the signal occupies. While the resulting signal sounds void of base response, its narrow 2.6 kHz of bandwidth really punches through background noise at the other end of the circuit. If you plan to do a lot of contesting with a huge antenna system, consider a step up to a $1500 high frequency base station with selectable transmit voice bandwidths.
ANSWER A

G8A06 What is one advantage of carrier suppression in a single-sideband phone transmission?

A. Audio fidelity is improved
B. Greater modulation percentage is obtainable with lower distortion
C. More transmitter power can be put into the remaining sideband
D. Simpler receiving equipment can be used

Carrier suppression allows *additional power* to be placed *in the sideband*.
ANSWER C

G4D01 What is the reason for using a properly adjusted speech processor with a single sideband phone transmitter?

A. It reduces average transmitter power requirements
B. It reduces unwanted noise pickup from the microphone
C. It improves voice-frequency fidelity
D. It improves signal intelligibility at the receiver

Properly adjusting your transmitter's speech processor will *improve the intelligibility* of your signal allowing it to be better heard at the receiving station.
ANSWER D

G4D02 Which of the following describes how a speech processor affects a transmitted single sideband signal?

A. It increases the peak power
B. It increases the average power
C. It reduces harmonic distortion
D. It reduces intermodulation distortion

Turning on your transceiver's *speech processor* will not increase PEP output power if you are 100 percent modulated. It will only *increase the average power*, and many times makes your signal sound "too hot" for comfort. Stay off that speech processor button unless it's absolutely necessary. **ANSWER B**

G8A09 What control is typically adjusted for proper ALC setting on an amateur single sideband transceiver?

A. The RF Clipping Level
B. Audio or microphone gain
C. Antenna inductance or capacitance
D. Attenuator Level

I normally run my *microphone gain control* on the SSB transceiver straight up and down in the 12:00 o'clock position. This allows a slight movement of the ALC meter on modulation peaks. As long as the meter stays within the ALC region marked on the meter, everything is fine. **ANSWER B**

G8A08 What happens to the signal of an over-modulated single-sideband phone transmitter?

A. It becomes louder with no other effects
B. It occupies less bandwidth with poor high frequency response
C. It has higher fidelity and improved signal to noise ratio
D. It becomes distorted and occupies more bandwidth

If you turn your transceiver's microphone gain too high, or engage the speech processor and turn it up too high, *your signal will become distorted and* will *occupy more bandwidth*. **ANSWER D**

G4D03 Which of the following can be the result of an incorrectly adjusted speech processor?

A. Distorted speech
B. Splatter
C. Excessive background pickup
D. All of these answers are correct

Most newer high frequency ham transceivers have a speech processor button that will either turn on processing or turn it off. TURN THE SPEECH PROCESSOR OFF! Speech processing brings in *excessive background pickup*, and may cause your signal to *sound distorted* and may cause it also to *splatter* to adjacent frequencies. Leave your speech processor turned off as a new General Class ham!
ANSWER D

G8A10 What is meant by flat-topping of a single-sideband phone transmission?

A. Signal distortion caused by insufficient collector current
B. The transmitter's automatic level control is properly adjusted
C. Signal distortion caused by excessive drive
D. The transmitter's carrier is properly suppressed

If you turn the microphone *gain too high* on SSB, that brand new rig will *sound distorted*. The signal waveform shown on an oscilloscope has the top clipped off so it has a *flat top*. **ANSWER C**

Elmer's Oscilloscope Waveform Showing "Flattopping"

G4A11 What type of transmitter performance does a two-tone test analyze?
A. Linearity
B. Carrier and undesired sideband suppression
C. Percentage of frequency modulation
D. Percentage of carrier phase shift

The faithful reproduction of your voice over single-sideband depends on equipment with excellent *linearity*. With new digital signal processing techniques on modulation, plus a new family of after-market, high-performance microphones, using the 2-tone test to view the wave forms on an oscilloscope is a great way to check for transmitter linearity. But you know, even though the 2-tone test is your best correct answer for linearity in the transmitter, I always like to ask fellow hams simply how I sound coming over their radio. No oscilloscope needed! **ANSWER A**

G4A02 Which of the following instruments may be used to measure the output of a single-sideband transmitter when performing a two-tone test of amplitude linearity?
A. An audio distortion analyzer
B. An oscilloscope
C. A directional wattmeter
D. A high impedance audio voltmeter

We use *two audio tones* to test for proper linearity while viewing the tones on an *oscilloscope*. These pure tones, not harmonically related, will give you a stable picture on the scope if your transmitter amplifier has proper linearity. **ANSWER B**

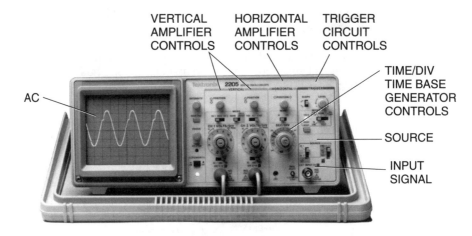

A Typical Oscilloscope Showing an AC Waveform
Source: *Basic Electronics* © 1994, Master Publishing, Inc., Niles, Illinois

G4B01 What item of test equipment contains horizontal and vertical channel amplifiers?

A. An ohmmeter
B. A signal generator
C. An ammeter
D. An oscilloscope

An oscilloscope is your best piece of test equipment if you are a technical amateur operator. But it takes skill to work a "scope," so don't buy one unless you know how to use it. The *oscilloscope* has horizontal- and vertical-channel amplifiers.
ANSWER D

G4B05 Which of the following is the best instrument to use to check the keying waveform of a CW transmitter?

A. A monitoring oscilloscope   C. A sidetone monitor
B. A field-strength meter   D. A wavemeter

The oscilloscope is your best instrument to check for transmit signal quality. When magazine editors review the waveform of a CW signal or a two-tone test, they usually show photographs of the *oscilloscope* display. **ANSWER A**

G4B06 What signal source is connected to the vertical input of a monitoring oscilloscope when checking the quality of a transmitted signal?

A. The local oscillator of the transmitter
B. The audio input of the transmitter
C. The transmitter balanced mixer output
D. The attenuated RF output of the transmitter

We take the *attenuated RF output* of the transmitter and couple it *to the vertical input* of a monitoring oscilloscope to check the quality of the transmitted signal. **ANSWER D**

G4B02 Which of the following is an advantage of an oscilloscope versus a digital voltmeter?

A. An oscilloscope uses less power
B. Complex impedances can be easily measured
C. Input impedance is much lower
D. Complex waveforms can be measured

When you whistle into your SSB microphone and look at the wave forms on an oscilloscope, you will see that only an *oscilloscope can* truly *represent complex voice wave forms* that can be precisely measured. **ANSWER D**

G4A12 What type of signals are used to conduct a two-tone test?

A. Two audio signals of the same frequency shifted 90-degrees
B. Two non-harmonically related audio signals
C. Two swept frequency tones
D. Two audio frequency range square wave signals of equal amplitude

More technical hams who regularly service their own equipment have developed *2 non-harmonically related audio tones* in a little "warbler" box that may either be plugged into the transmitter mike input, or the mike held up to the little warbler speaker. The 2 tones must not be harmonically related in order to provide the best test. **ANSWER B**

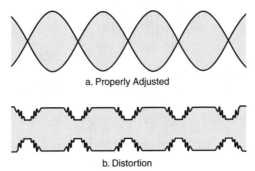

a. Properly Adjusted

b. Distortion

Two-Tone Test

G4A10 What is the reason for neutralizing the final amplifier stage of a transmitter?
A. To limit the modulation index
B. To eliminate self oscillations
C. To cut off the final amplifier during standby periods
D. To keep the carrier on frequency

After new tubes have been installed in a powerful amplifier, follow the instruction manual steps for *neutralization*, which will *eliminate* the possibility of *self-oscillation*. **ANSWER B**

G8A03 What is the name of the process which changes the frequency of an RF wave to convey information?
A. Frequency convolution   C. Frequency conversion
B. Frequency transformation   D. Frequency modulation

A type of modulation that changes the frequency of an RF wave is called *frequency modulation (FM)*. **ANSWER D**

G8A11 What happens to the RF carrier signal when a modulating audio signal is applied to an FM transmitter?
A. The carrier frequency changes proportionally to the instantaneous amplitude of the modulating signal
B. The carrier frequency changes proportionally to the amplitude and frequency of the modulating signal
C. The carrier amplitude changes proportionally to the instantaneous frequency of the modulating signal
D. The carrier phase changes proportionally to the instantaneous amplitude of the modulating signal

When you get your new General Class license, you STILL will be working with the gang on 2 meters and 440 MHz FM. The way your *FM* transceiver *changes* your voice to a signal on the air is the carrier *frequency* changes proportionately TO THE INSTANTANEOUS AMPLITUDE of the modulating signal. Now say 5 times, out loud, "Frequency changes instantaneous amplitude," and then again, "Frequency changes instantaneous amplitude," and then again, *"Frequency changes instantaneous amplitude."* Got it? **ANSWER A**

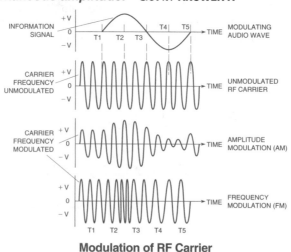

Modulation of RF Carrier

G8B04 What is the name of the stage in a VHF FM transmitter that selects a harmonic of an HF signal to reach the desired operating frequency?
A. Mixer
B. Reactance modulator
C. Pre-emphasis network
D. Multiplier

Inside a VHF FM transmitter is a low-power oscillator that operates at HF levels. It's the job of the *multiplier* to *select a harmonic* of this signal to produce the desired operating frequency. **ANSWER D**

G8B06 What is the total bandwidth of an FM-phone transmission having a 5 kHz deviation and a 3 kHz modulating frequency?
A. 3 kHz
B. 5 kHz
C. 8 kHz
D. 16 kHz

You can calculate this answer by multiplying 2 times the sum of the deviation and highest audio modulating frequency. The deviation is 5 kHz plus 3 kHz of audio, which gives a sum of 8 kHz. Two times 8 kHz equals *16 kHz*. This would be the *total bandwidth* of the FM phone transmission. **ANSWER D**

G8B07 What is the frequency deviation for a 12.21-MHz reactance-modulated oscillator in a 5-kHz deviation, 146.52-MHz FM phone transmitter?
A. 101.75 Hz
B. 416.7 Hz
C. 5 kHz
D. 60 kHz

This is an easy ratio problem. First, let's determine the frequency multiplication factor of the transmitter. Divide the 12.21 oscillator frequency into the output at 146.52. This gives us a multiplication factor of 12. If there is 5 kHz deviation at the transmitter, 1/12th deviation at the oscillator input is 12 divided into 5000 Hz, with an answer of *416.66 Hz*. **ANSWER B**

G8A04 What emission is produced by a reactance modulator connected to an RF power amplifier?
A. Multiplex modulation
B. Phase modulation
C. Amplitude modulation
D. Pulse modulation

A *reactance modulator* produces *phase modulation*, which is used for phase modulation and frequency modulation. **ANSWER B**

G8A02 What is the name of the process that changes the phase angle of an RF wave to convey information?
A. Phase convolution
B. Phase modulation
C. Angle convolution
D. Radian Inversion

A type of modulation that changes the phase of an RF wave is called *phase modulation (PM)*. **ANSWER B**

G8B10 When transmitting a data mode signal, why is it important to know the duty cycle of the mode you are using?
A. To aid in tuning your transmitter
B. To prevent damage to your transmitter's final output stage
C. To allow time for the other station to break in during a transmission
D. All of these choices are correct

Hooking your high-frequency ham transceiver up to the modern computer opens up a whole new world of receiving and sending data signals. Receiving is almost a direct connection to your computer through the sound card, but sending data may require an external modem. Sending data with on-and-off handshake modes like

PACTOR II or G-TOR cycles your new General Class transceiver to transmit and receive for a duty cycle (on transmit) of perhaps 50 percent. But other modes like PSK31 and MFSK16 are a constant key-down data stream without interruption. Now your transmit duty cycle is 100%, and the heat sink on the back of your transceiver is going to get roasty-toasty. To keep your equipment from going into meltdown in the data modes for transmitting, consider an external fan plus reducing power output to the point where the rear heat sinks won't fry eggs. If you get the heat sinks so warm that you can't touch them, *you could damage the transmitter final output stage* and that will lead to a mighty expensive repair. **ANSWER B**

G7B13 How is the efficiency of an RF power amplifier determined?
 A. Divide the DC input power by the DC output power
 B. Divide the RF output power by the DC input power
 C. Multiply the RF input power by the reciprocal of the RF output power
 D. Add the RF input power to the DC output power

You have your brand new high frequency transceiver hooked up to a watt meter and a dummy load. The sales brochure indicates that new rig offers 250 watts input. Yet, when you key the microphone, and look at the watt meter into the dummy load and say the word "FOOOUUUURRRRR," the needle barely goes to 100 watts output. Hey, what's going on here? Power input to the final transistor block yields about 50 % efficiency or slightly less using voice. Some of the power goes up as heat, a little bit of power may drive other stages, and the resulting OUTPUT POWER will always be much less than INPUT POWER. To calculate the efficiency of your new high frequency transceiver, you *divide the RF output power by the DC input power*. The RF output power is easily seen on the watt meter, and if you can whistle in the mike and get 100 watts out, your rig is working as it should. If you want to get scientific and calculate input power, you will need to calculate voltage times transistor block current. But just remember the formula: efficiency = power output divided by power input. And REMEMBER, high frequency output power as measured with an inline watt meter for voice will dance around 60 watts, and this is plenty to work the world with your new General Class privileges. **ANSWER B**

This linear power amp can boost the 1- to 5-watt output of a hand-held up to 30 watts. The FCC-certified amp meets stringent spurious emission standards.

G5B06 What is the output PEP from a transmitter if an oscilloscope measures 200 volts peak-to-peak across a 50-ohm dummy load connected to the transmitter output?

    A. 1.4 watts   C. 353.5 watts

    B. 100 watts   D. 400 watts

Peak envelope power is abbreviated PEP. Multiply the peak envelope voltage (PEV) by the RMS value (0.707), multiply this by itself (square the result), and then divide the squared quantity by the resistance of the load. Here is the formula:

    PEP = (peak voltage x 0.707)^2 ÷ resistance of the load

Watch out – this question reads 200 volts peak-to-peak. Peak voltage is one-half of the peak-to-peak value, or 100. Multiply 100 x 0.707 for 70.7. Multiply 70.7 by 70.7 to square the result for (70.7)^2 then divide by 50. Here are the calculator key strokes: Clear 100 x 0.707 x 70.7 ÷ 50 = 99.97, which rounds off to *100 watts*.
ANSWER B

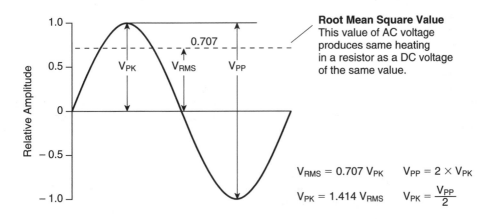

RMS (V$_{RMS}$), Peak, (V$_{PK}$), and Peak-to-Peak (V^{pp}) Voltage

---

G5B11 What is the ratio of peak envelope power to average power for an unmodulated carrier?

    A. 0.707   C. 1.414

    B. 1.00   D. 2.00

An unmodulated carrier is simultaneously peak and average. So *the ratio of peak envelope power to average power on a steady unmodulated carrier is simply 1.* **ANSWER B**

G5B14 What is the output PEP from a transmitter if an oscilloscope measures 500 volts peak-to-peak across a 50-ohm resistor connected to the transmitter output?

    A. 8.75 watts   C. 2500 watts

    B. 625 watts   D. 5000 watts

Again take half the peak-to-peak voltage to obtain the peak voltage. Multiply 250 x 0.707, square the result (176.75) by multiplying it by itself, and divide by 50. Here are the calculator key strokes: Clear, Clear 250 x 0.707 x 176.75 ÷ 50= 624.81, which rounds off to *625 watts*. **ANSWER B**

G5B15 What is the output PEP of an unmodulated carrier if an average reading wattmeter connected to the transmitter output indicates 1060 watts?
- A. 530 watts
- B. 1060 watts
- C. 1500 watts
- D. 2120 watts

Remember a steady carrier illustrates both peak power as well as average power. If your watt meter reads *1060 watts* average, this is the same power as peak envelope power. **ANSWER B**

G7A10 What is an advantage of a crystal controlled transmitter?
- A. Stable output frequency
- B. Excellent modulation clarity
- C. Ease of switching between bands
- D. Ease of changing frequency

Today's modern high frequency transceiver still continues to use a quartz crystal reference oscillator. Thanks to PLL, you are no longer "rock bound" like the old days of crystals. The importance of the *quartz crystal is stable output frequency* generation. Transceivers that incorporate a TCO (temperature compensated oscillator) are preferred for digital operation because the crystal itself is in a tiny oven to keep it rock steady on frequency. **ANSWER A**

G5B01 A two-times increase or decrease in power results in a change of how many dB?
- A. 2 dB
- B. 3 dB
- C. 6 dB
- D. 12 dB

The decibel is used to describe a change in power levels. It is a measure of the ratio of power output to power input. *A two-times increase results in a change of 3 dB.* **ANSWER B**

If $dB = 10 \log_{10} \dfrac{P_1}{P_2}$

then what power ratio is 20 dB?

$20 = 10 \log_{10} \dfrac{P_1}{P_2}$

$\dfrac{20}{10} = \log_{10} \dfrac{P_1}{P_2}$

$2 = \log_{10} \dfrac{P_1}{P_2}$

Remember: logarithm of a number is the exponent to which the base must be raised to get the number.

$\therefore 10^2 = \dfrac{P_1}{P_2}$

$100 = \dfrac{P_1}{P_2}$

Or $P_1 = 100\, P_2$

20 dB means P_1 is 100 times P_2

dB	$\dfrac{P_1}{P_2}$
3	2
6	4
10	10
20	100
30	1000
40	10000
50	10^5
60	10^6

Definition of a Decibel

Source: *The Technology Dictionary*,© 1987 Master Publishing, Inc., Niles, IL

G4B09 How much must the power output of a transmitter be raised to change the "S" meter reading on a distant receiver from S8 to S9?
A. Approximately 2 times   C. Approximately 4 times
B. Approximately 3 times   D. Approximately 5 times

Seeing an S-meter change from S8 to S9 is an increase of a single S unit. *One S unit is 6 dB*, and *6 dB is a 4-times change*. **ANSWER C**
Here is how the dB system for power works:

> 0 dB = 0 times change
>
> 3 dB = 2 times change
>
> 6 dB = 4 times change
>
> 9 dB = 8 times change
>
> 10 dB = 10 times change

G4D05 How does an S-meter reading of 20 db over S-9 compare to an S-9 signal, assuming a properly calibrated S meter?
A. It is 10 times weaker   C. It is 20 times stronger
B. It is 20 times weaker   D. It is 100 times stronger

An S meter reading of 20 db over S-9, compared to just an S-9 signal, illustrates the signal is *100 times stronger*. **ANSWER D**

G7B11 For which of the following modes is a Class C power stage appropriate for amplifying a modulated signal?
A. SSB   C. AM
B. CW   D. All of these answers are correct

The *Class C amplifier* is ideal for non-linear amplification of *Morse code (CW)* plus many digital emissions. Varying amplitude emissions like SSB or AM would not sound great if the amplifier is operating in the Class C region. But for CW, it will work great. **ANSWER B**

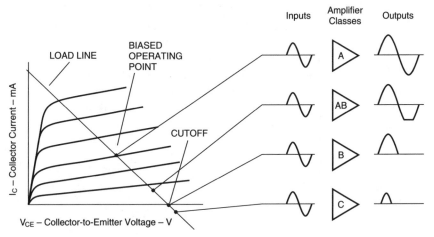

Various Classes of Transistorized Amplifiers

G7B12 Which of the following is an advantage of a Class C amplifier?

A. High efficiency
B. Linear operation
C. No need for tuned circuits
D. All of these answers are correct

Yes, Grandpa, I will learn the code! CW (Morse code) will always continue to be a popular mode for ham operators, even though the code test has been completely eliminated for your upcoming General Class exam. A small CW-only transceiver, operating *Class C* offers *high efficiency* with little current being consumed from the battery in between dots and dashes. Class C, high efficiency. Class A, low distortion. **ANSWER A**

G7B10 Which of the following is a characteristic of a Class A amplifier?

A. Low standby power
B. High Efficiency
C. No need for bias
D. Low distortion

The *Class A* amplifier in your new high frequency transceiver is many times found within the microphone input stage, where *low distortion* is an absolute requirement. Class A amplifiers are continuously drawing a small amount of current to provide the utmost linear operation. **ANSWER D**

G7B14 Which of the following describes a linear amplifier?

A. Any RF power amplifier used in conjunction with an amateur transceiver
B. An amplifier whose output preserves the input waveform
C. A Class C high efficiency amplifier
D. An amplifier used as a frequency multiplier

Don't run right out and buy a $1000 linear amplifier to boost the power output on your $700-$900 high frequency transceiver. *Linear amplifiers* will indeed pump up your *output* watts *exactly as* they appear to the amplifier *input*, but save linear operation only after you have been on the air for at least a year and plan to put up a 100 ft tower and a major Yagi antenna. It is unwise to run any linear amplifier on high frequency mobile or on a simple home antenna until after you see that 100 powerful watts can work all around the world. **ANSWER B**

Elmer Point: *Don't run out and buy a linear amplifier. The 100-watt output from your HF ham transceiver is plenty powerful enough to work the world. If you want to blast your signal out stronger, go for a directional antenna. The directional antenna also will increase incoming received signal strength – something that linear amplifier can't do. Linear amps are for those hams who have been on the air for at least a year, and who put up a 60 foot tower with a big directional beam on top. What the amplifier will do is increase their ability to bust through a pileup. Pileups occur when numerous hams are simultaneously trying to get their call signs through to a rare DX station during a contest. When you're just getting started on the General Class airwaves, steer clear of contest operation where pileups take place.*

BIG MAMA
750 / 1500
WATTS
POWER AMPLIFIER

G4A07 What is the correct adjustment for the "Load" or "Coupling" control of a vacuum tube RF power amplifier?

    A. Minimum SWR on the antenna
    B. Minimum plate current without exceeding maximum allowable grid current
    C. Highest plate voltage while minimizing grid current
    D. Maximum power output without exceeding maximum allowable plate current

On the load control on that old power amp, seasoned hams will tell you to "tune for maximum smoke." Well, not really – but tune for a *maximum power output* reading, making sure you *do not exceed 1500 watts output*. Most older amps will do 1,000 watts and not much more. **ANSWER D**

G4A06 What reading on the plate current meter of a vacuum tube RF power amplifier indicates correct adjustment of the plate tuning control?

    A. A pronounced peak
    B. A pronounced dip
    C. No change will be observed
    D. A slow, rhythmic oscillation

Vacuum tube linear amplifiers are always a welcome ham radio "accessory" for the shack – especially if your Granddad gives you his old linear amplifier! When hooked up to a properly-tuned antenna, the plate tuning control will usually show a *pronounced dip* on the amplifier's big meter. If you don't see a pronounced dip, time to see what went wrong with your antenna system. The more resonant your tuned antenna is, the more pronounced the dip in your plate current meter reading. **ANSWER B**

G4A09 What does a neutralizing circuit do in an RF amplifier?

    A. It controls differential gain
    B. It cancels the effects of positive feedback
    C. It eliminates AC hum from the power supply
    D. It reduces incidental grid modulation

Neutralization of an RF amplifier *cancels* the effects of *positive feedback*. When an amplifier goes into positive feedback, the grid current increases to a full deflection, the amplifier goes into oscillation, and unless you un-key the mike quickly, the output tube will go into meltdown in short order. **ANSWER B**

G4A08 Which of the following techniques is used to neutralize an RF amplifier?

    A. Feed-forward compensation
    B. Feed-forward cancellation
    C. Negative feedback
    D. Positive feedback

When we neutralize an amplifier, we take a small amount of the output signal, shift it 180 degrees, and feed it back into the input. The feedback signal works against the input signal so it is called *negative feedback*. **ANSWER C**

Your Receiver

G7A11 What is the simplest combination of stages that can be combined to implement a superheterodyne receiver?

A. RF amplifier, detector, audio amplifier
B. RF amplifier, mixer, if amplifier
C. HF oscillator, mixer, detector
D. HF oscillator, product detector, audio amplifier

The early (before you were born) ham radio high frequency receiver operated as TRF – tuned radio frequency. The receiver was nice and sensitive, but not all that stable. Today, our modern superheterodyne ham radio receiver uses intermediate frequency amplifiers with deep narrow frequency filters that offer plenty of gain, excellent frequency stability, and menu-optional selectivity and sensitivity. The term "superheterodyne" indicates a receiver with local oscillators and mixers which recover an incoming signal clean of other signals just above and below the desired signal. New ham transceivers now allow the operator to customize the intermediate frequency band pass filters for just the right amount of bandwidth to clearly hear the incoming weak signal. The simplest combination of stages to describe a superheterodyne receiver would be the *HF oscillator*, the *mixer* stage, and the *detector*. **ANSWER C**

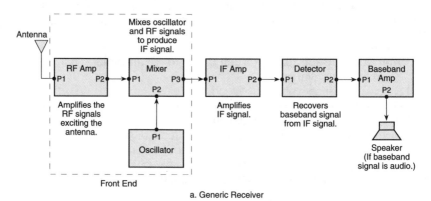

a. Generic Receiver

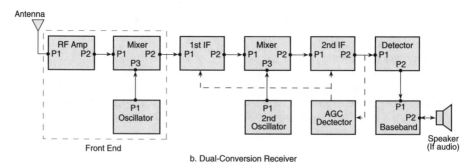

b. Dual-Conversion Receiver

Block Diagrams of Generic and Dual-Conversion AM Receivers
Source: Basic Communications Electronics,
© 1999 Master Publishing, Inc., Niles, IL

G7A08 What circuit is used to process signals from the RF amplifier and local oscillator and send the result to the IF filter in a superheterodyne receiver?

A. Balanced modulator   C. Mixer
B. IF amplifier   D. Detector

In an SSB receiver section, it is the *mixer* that processes signals from the RF amplifier and the local oscillator. **ANSWER C**

G7A09 What circuit is used to process signals from the IF amplifier and BFO and send the result to the AF amplifier in a single-sideband phone superheterodyne receiver?

A. RF oscillator   C. Balanced modulator
B. IF filter   D. Product detector

In an SSB receiver it is the *product detector* that processes the signal from the IF amplifier and the beat frequency oscillator. The signal then goes on to the audio frequency amplifier. **ANSWER D**

G8B12 What is another term for the mixing of two RF signals?

A. Heterodyning   C. Cancellation
B. Synthesizing   D. Multiplying

When a ham is tuning around the worldwide band and is listening to a juicy conversation, and then another signal comes right on top of it, they sometimes refer to this as a "heterodyne." This was a major problem in the early days of AM CB radio. But believe it or not, the inside of your equipment uses *heterodyning* for the positive purpose of *mixing 2 RF signals* near the IF stage. **ANSWER A**

G8B03 What stage in a transmitter would change a 5.3 MHz input signal to 14.3 MHz?

A. A mixer   C. A frequency multiplier
B. A beat frequency oscillator   D. A linear translator

It's the job of the *mixer* to change a 5.3 MHz input signal up to 14.3 MHz by mixing in an oscillator frequency. **ANSWER A**

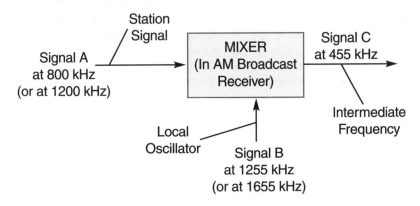

Block Diagram of an AM Broadcast Receiver Mixer
Source: *Basic Communications Electronics,*
© 1999 Master Publishing, Inc., Niles, IL

G8B01 What receiver stage combines a 14.250 MHz input signal with a 13.795 MHz oscillator signal to produce a 455 kHz intermediate frequency (IF) signal?
A. Mixer   C. VFO
B. BFO   D. Multiplier
When you see that word "combines," think of the *mixer* section of a receiver stage.
ANSWER A

G8B02 If a receiver mixes a 13.800 MHz VFO with a 14.255 MHz received signal to produce a 455 kHz intermediate frequency (IF) signal, what type of interference will a 13.345 MHz signal produce in the receiver?
A. Local oscillator   C. Mixer interference
B. Image response   D. Intermediate interference
In strong signal areas where there may be local transmissions coming in from shortwave stations outside of normal ham band limits, an interference called *"image response"* may develop at the sum and difference of your intermediate frequency (IF) signal. 13.800 MHz minus 455 kHz is 13.345 MHz. **ANSWER B**

G4A04 Which of the following is an advantage of a receiver IF filter created with a DSP as compared to an analog filter?
A. A wide range of filter bandwidths and shapes can be created
B. Fewer digital components are required
C. Mixing products are greatly reduced
D. The DSP filter is much more effective at VHF frequencies
The new, modern high frequency transceiver usually includes *digital signal processing*. This would allow *multiple filter bandwidth settings*, along with bandwidth shape for "soft" or "hard" response. This wide range of filter combinations – all digital – really helps subtract background noise. **ANSWER A**

Elmer Point: *If you start out with a used, older, HF radio, you can buy an audio DSP speaker system for about $125. You'll be amazed at how the DSP noise subtraction circuit improves the sounds coming from the radio. But better yet, buy a new HF transceiver with DSP filtering in the IF and now you'll really hear a big difference of cleaned-up reception. DSP is not magic, however. The best way to improve reception is to search out whatever it is around your house that is generating all the noise and get it shut down when you begin to operate with those distant, rare, weak-signal stations. Fans and florescent lights are huge noise generators.*

G4A03 Which of the following is needed for a DSP IF filter?
A. An Analog to Digital Converter
B. Digital to Analog Converter
C. A Digital Processor Chip
D. All of the these answers are correct
Inside a DSP integrated circuit is a digital processor that converts analog to digital, a noise subtraction stage, and then digital to analog converter. So, *all of these answers* are correct in the modern DSP IF filter. **ANSWER D**

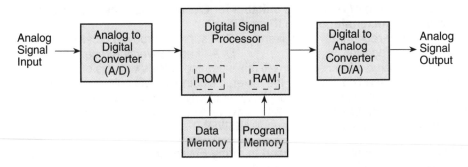

Block Diagram of a Basic Digital Signal Processing (DSP) System
Source: *Basic Communications Electronics,*
© 1999 Master Publishing, Inc., Niles, IL

G4A01 Which of the following is one use for a DSP in an amateur station?
A. To provide adequate grounding
B. To remove noise from received signals
C. To increase antenna gain
D. To increase antenna bandwidth

Digital signal processing (DSP) is found in almost all new high frequency amateur radio transceivers. Older HF sets don't have DSP, and this is why I always recommend buying a NEW HF transceiver. New HF equipment is available for under $550, 100 watts out, 100 channels of memory, DSP included! *DSP* is a great way to *remove noise* from received weak signals. **ANSWER B**

G4A05 How is DSP filtering accomplished?
A. By using direct signal phasing
B. By converting the signal from analog to digital and using digital processing
C. By up-converting the signal to VHF
D. By converting the signal from digital to analog and taking the difference of mixing products

The input of a *DSP* circuit *takes* an *analog* signal, *converts* it *to* a *digital* signal where the *noise subtraction* takes place, and *restores it to an analog* signal so we can hear it again! **ANSWER B**

G4A13 Which of the following performs automatic notching of interfering carriers?
A. Band pass tuning   C. Balanced mixing
B. A DSP filter   D. A noise limiter

Always buy new ham gear to get started on your new high frequency privileges. New gear incorporates a circuit called "ANF" on the front panel – automatic notch filter. This is handy on 40 meters where there is a lot of foreign broadcast carrier noise that will drive you crazy without the automatic notch filter. *"ANF" uses a DSP filter to* eliminate the steady tone and magically the signal you are trying to understand will now come in more clearly. **ANSWER B**

G7A12 What type of receiver is suitable for CW and SSB reception but does not require a mixer stage or an IF amplifier?
A. A super-regenerative receiver   C. A superheterodyne receiver
B. A TRF receiver   D. A direct conversion receiver
The *direct conversion receiver* will mix incoming signals with the output of the local oscillator, feeding them into the mixer, including the LO output. While less sophisticated than the superhet receiver, sensitivity is good but individual station selectivity is not so great. This is why I always recommend BUY NEW when considering high frequency equipment. **ANSWER D**

G7A13 What type of circuit is used in many FM receivers to convert signals coming from the IF amplifier to audio?
A. Product detector   C. Mixer
B. Phase inverter   D. Discriminator
In a frequency modulation receiver, signals coming from the intermediate frequency amplifier are fed into the *discriminator*, which acts as a frequency to voltage conversion stage, sending the resulting "decoded" signal on to the audio amplifier. **ANSWER D**

G4D06 Where is an S-meter generally found?
A. In a receiver   C. In a transmitter
B. In a SWR bridge   D. In a conductance bridge
The S meter on your new HF transceiver ties directly in to the *receiver* section of the radio. Although the large meter movement may illustrate other radio parameters, when it is in the "S-meter" mode, it is tied in to the receiver's signal strength circuitry. **ANSWER A**

G4D04 What does an S-meter measure?
A. Conductance   C. Received signal strength
B. Impedance   D. Transmitter power output
High frequency ham transceivers all incorporate an S-meter. *S-meters measure received signal strength*, and might also be used to indicate a properly-working antenna system. On 40 meters, you should see an approximately S-2 indication of normal background noise. If you see a higher reading of background noise, this is indeed an indication that your antenna is performing properly. However, if your S meter doesn't budge on 40 meter noise, something is probably wrong with your antenna system. **ANSWER C**

G4B03 How would a signal tracer normally be used?
A. To identify the source of radio transmissions
B. To make exact drawings of signal waveforms
C. To show standing wave patterns on open-wire feedlines
D. To identify an inoperative stage in a receiver
A signal tracer provides a source signal to the point in the circuit chosen by you. With normal receiver volume, you can usually hear the tone of the signal tracer coming through. Start at the speaker circuit and work backwards, stage by stage, until the tone abruptly disappears. This will allow you to *detect a stage that may have a problem*. **ANSWER D**

Website Resources

▼ IF YOU'RE LOOKING FOR	▼ THEN VISIT
Ham Equipment Reviews	www.eham.net
HRO – Ham Radio Outlet	www.hamradio.com
AES – Amateur Radio Supply	www.aesham.com
Largest Ham Accessory Catalog	www.mfjenterprisses.com
Advanced Specialties	www.advancedspecialties.net
Alltronics	www.alltronics.com
Amateur Accessories	www.amateuraccessories.com
Amateur Radio Toy Store, Inc.	www.amateur-radio-toy-store.com
Asscociated Radio	www.associatedradio.com
Austin Amateur Radio	www.aaradio.com
B&H Sales	www.hamradiocenter.com
Bill Rouch Marketing	www.hamneeds.com
Bob Elder, KG4ENA	www.n7wwk.com
Cedar City Sales	www.cedarcitysales.com
Central Utah Electronics Supply	www.electronicspro.com
Communications Products	www.commproducts.net
DBJ Radio & Electronics	www.dbjre.com
GigaParts, Inc.	www.gigaparts.com
HamStop.com	www.hamstop.com
Houston Amateur Radio Supply	www.texasparadise.com/hars
Iowa Radio Supply Co., Inc	www.irsupply.com
Jubilee Enterprises	www.shopjubilee.com
Jun's Electronics	www.hamcity.com
K1CRA Radio Webstore	www.k1cra.com
K-Comm, Inc. – The Ham Store	www.kcomm.biz
KJI Electronics, Inc.	www.kjielectronics.com
Lentini Communications, Inc.	www.lentinicomm.com
R & L Electronics	www.randl.com
Rad-Comm Radio	www.radcomm.bizland.com/rad-comm
Radio City	www.radioinc.com
Rayfield Communications	www.rayfield.net
The Ham Station	www.hamstation.com
Universal Radio Inc	www.universal-radio.com
Waypoint Crusing Solutions	www.waypoints.com
WBOW, Inc.	www.wb0w.com

Note: This list includes many ham radio dealers where you can purchase your HF radio and accessories

G7B07 What are the basic components of virtually all oscillators?
    A. An amplifier and a divider
    B. A frequency multiplier and a mixer
    C. A circulator and a filter operating in a feed-forward loop
    D. A filter and an amplifier operating in a feedback loop

To keep an oscillator "in motion," a tuned circuit at one specific frequency will provide the necessary *FEEDBACK LOOP*. Just like keeping Grandpa swinging back and forth on his brand new senior citizen swing set, a small amount of FEEDBACK continuously keeps him in motion. **ANSWER D**

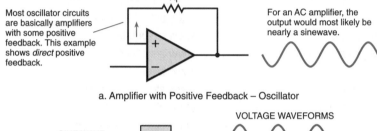

Most oscillator circuits are basically amplifiers with some positive feedback. This example shows *direct* positive feedback.

For an AC amplifier, the output would most likely be nearly a sinewave.

a. Amplifier with Positive Feedback – Oscillator

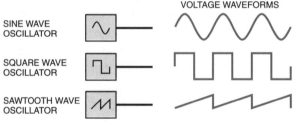

VOLTAGE WAVEFORMS

SINE WAVE OSCILLATOR

SQUARE WAVE OSCILLATOR

SAWTOOTH WAVE OSCILLATOR

b. Oscillator Waveforms

An oscillator is basically an amplifier with positive feedback from output to input.

Source: *Basic Electronics* © 1994, Master Publishing, Inc., Niles, Illinois

G7B08 What determines the frequency of an RC oscillator?
    A. The ratio of the capacitors in the feedback loop
    B. The value of the inductor in the tank circuit
    C. The phase shift of the RC feedback circuit
    D. The gain of the amplifier

An *RC oscillator* uses a combination of resistance and capacitance (RC) to provide a *phase shift in the positive feedback loop*. RC oscillators are usually found in the audio output section of your new high frequency transceiver. **ANSWER C**

G7B09 What determines the frequency of an LC oscillator?
    A. The number of stages in the counter
    B. The number of stages in the divider
    C. The inductance and capacitance in the tank circuit
    D. The time delay of the lag circuit

An LC oscillator is found in the radio frequency circuits of your new HF transceiver, and consists of a tuned *inductance and capacitance* resonant (LC) circuit, found *in the tank circuit* of that new "Gee Whiz" high frequency transceiver. **ANSWER C**

G6B10 Which element of a triode vacuum tube is used to regulate the flow of electrons between cathode and plate?
A. Control grid
B. Heater
C. Screen Grid
D. Suppressor grid
In the triode vacuum tube, the *CONTROL GRID* acts like a variable valve to regulate the flow of electrons between the cathode and plate.
ANSWER A

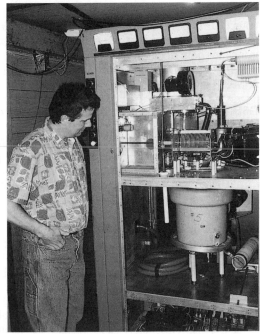

High-power vacuum tubes require neutralization.

G6B12 What is the primary purpose of a screen grid in a vacuum tube?
A. To reduce grid-to-plate capacitance
B. To increase efficiency
C. To increase the high frequency response
D. To decrease plate resistance
Yes, hams still run rigs with vacuum tubes. A diode tube has an anode and a cathode that may incorporate direct or indirect heating. The diode tube is normally used in power supplies. The triode tube incorporates a control grid and, like a variable amplifier, the control grid acts as a valve. Next is the tetrode tube, and it adds a *screen grid*, which will *reduce grid-to-plate capacitance* and collects secondary emissions from the plate and will slightly increase screen grid current.
ANSWER A

G6B11 Which of the following solid state devices is most like a vacuum tube in its general characteristics?
A. A bipolar transistor   C. A tunnel diode
B. An FET   D. A varistor
The *field effect transistor (FET)* has high input impedance, and has a gate, drain and source. The gate is similar to the control grid of a tube, with the gate generating an electric field that increases or decreases the amount of current flow. **ANSWER B**

G6B08 Why is it often necessary to insulate the case of a large power transistor?
A. To increase the beta of the transistor
B. To improve the power dissipation capability
C. To reduce stray capacitance
D. To avoid shorting the collector or drain voltage to ground

Be careful where you place the power supply that changes household power to 12 volts DC. Older transformer-type power supplies mount their power transistors on the large heat sinks that are exposed to the open air. If the power supply power transistor gets accidentally pushed up against something else that is ground, you will *short out the collector voltage* and likely destroy the power transistor. **ANSWER D**

G6B07 What are the stable operating points for a bipolar transistor that is used as a switch in a logic circuit?
- A. Its saturation and cut-off regions
- B. Its active region (between the cut-off and saturation regions)
- C. Between its peak and valley current points
- D. Between its enhancement and deletion modes

When the bipolar transistor reaches saturation, collector and emitter base junctions are forward-biased. When the bipolar transistor is used as a switch in a logic circuit, an extremely small change in collector-base voltage will cause a large change in collector current, allowing the transistor to switch between cut-off and saturation within collector current specifications to keep the transistor from self-destructing. It is in this *saturation and cut-off region* that the transistor will act as a switch. **ANSWER A**

G6B09 Which of the following describes the construction of a MOSFET?
- A. The gate is formed by a back-biased junction
- B. The gate is separated from the channel with a thin insulating layer
- C. The source is separated from the drain by a thin insulating later
- D. The source is formed by depositing metal on silicon

We will many times find the MOSFET transistor in the "front end" of your new modern HF transceiver. The *MOSFET* is a metal-oxide semiconductor field effect transistor, where the gate is separated from the channel with an *extremely THIN insulating layer*. Many times the manufacturer will install a gate-protective Zener diode, which prevents the gate insulation from being punctured by small static charges or excessive voltages such as a nearby lightening strike. **ANSWER B**

G6B03 What is the approximate junction threshold voltage of a germanium diode?

A. 0.1 volt	C. 0.7 volts
B. 0.3 volts	D. 1.0 volts

The *Germanium diode* is sometimes found in the detector stage of a receiver. Their low forward voltage drop is minimal, and the approximate junction *threshold voltage* of this diode is *0.3 volts*. These signal diodes are sensitive to burn out if you accidentally overheat them while soldering in a replacement. **ANSWER B**

G6B05 What is the approximate junction threshold voltage of a silicon diode?

A. 0.1 volt	C. 0.7 volts
B. 0.3 volts	D. 1.0 volts

We find *silicone diodes* in the rectifier section of your new transceiver's power supply. The silicone diode offers stable operation at high temperatures, a junction *threshold* voltage of about *0.7 volts*, and high reverse resistance along with long term reliability. **ANSWER C**

G6B06 Which of the following is an advantage of using a Schottky diode in an RF switching circuit as compared to a standard silicon diode?

A. Lower capacitance   C. Longer switching times
B. Lower inductance   D. Higher breakdown voltage

The *Schottky diode* is a fast switching point contact diode with *lower capacitance* offering faster switching capability. This is an important consideration in HF transceivers to handle many digital modes where fast switching time between transmit and receive is required. The lower the capacitance the greater is the advantage of the Schottky over a common silicone diode. **ANSWER A**

G6C07 What is one disadvantage of an incandescent indicator compared to a LED?

A. Low power consumption C. Long life
B. High speed   D. High power consumption

Long live the light emitting diode. These are favorites in ham radio transceivers. Luckily, they don't burn out like older incandescent lamps. The older *incandescent* indicator *lamps* also consume *high power levels*. Read carefully – this question asks for the DISadvantage of the incandescent light. **ANSWER D**

G6C08 How is an LED biased when emitting light?

A. Beyond cutoff   C. Reverse Biased
B. At the Zener voltage D. Forward Biased

The light emitting diode is always forward biased when it energizes. When current is passed through the PN junction, *forward biased*, the light emitting diode almost instantly turns on. **ANSWER D**

G6C09 Which of the following is a characteristic of a liquid crystal display?

A. It requires ambient or back lighting
B. It offers a wide dynamic range
C. It has a wide viewing angle
D. All of these choices are correct

Many of your new high frequency transceivers use a *liquid crystal display (LCD)*. They *require backlighting* at night, and the monochrome (black and white) liquid crystal displays achieve their own "brilliance" out in the sunshine. Color liquid crystal displays have dramatically improved readability in direct sunlight, and at night with backlighting, the color LED display really looks terrific! **ANSWER A**

Elmer Point: *If you plan to operate your radio in the field, make sure it offers a liquid crystal display. An amber background with black numbers is best for reading the display in bright sunlight. Color LCD displays are great at night, but they may require some shade during the day. Older HF transceivers may use LED or vacuum florescent display, and these are next to impossible to see when operating Field Day in a tent out in the sunshine. Go for amber LCD with black numbers for best viewing.*

G6C11 What is a microprocessor?
A. A low powered analog signal processor used as a microwave detector
B. A miniature computer on a single integrated circuit chip
C. A microwave detector, amplifier, and local oscillator on a chip
D. A low voltage amplifier used in a microwave transmitter modulator stage

Thanks to *microprocessors*, your new General Class high frequency transceiver works much like a *miniature computer* with everything on single integrated circuit chips. But just like your home computers, make sure your equipment has enough "breathing room." **ANSWER B**

G6C10 What is meant by the term MMIC?
A. Multi Megabyte Integrated Circuit
B. Monolithic Microwave Integrated Circuit
C. Military-specification Manufactured Integrated Circuit
D. Mode Modulated Integrated Circuit

The *monolithic microwave integrated circuit (MMIC)* is a favorite among microwave operators, up to 10 GHz. Up on "X" band, we use MMICs within our microwave equipment as fixed gain amplifiers with excellent signal to noise ratios. **ANSWER B**

Monolithic Microwave Integrated Circuit
Photo Courtesy of Hewlett Packard Co.

G6C06 Which type of integrated circuit is an operational amplifier?
A. Digital   C. Programmable
B. MMIC   D. Analog

We nickname the *operational amplifier an* "op amp", and this is an integrated circuit *analog device*. The op amp offers high gain over a large range of input frequencies, and is a direct-coupled differential amplifier whose characteristics are determined by components external to the amplifier unit. The "differential" wording refers to the op amp input design where the output is determined by the difference of voltages between the two inputs. The ideal op amp offers infinite input impedance, zero output impedance, infinite gain, and a flat analog frequency response. **ANSWER D**

G7B01 Which of the following describes a "flip-flop" circuit?

    A. A transmit-receive circuit
    B. A digital circuit with two stable states
    C. An RF limiter
    D. A voice-operated switch

Your new high frequency transceiver has many stages of "gee whiz" digital technology. An amazing circuit called the *"flip-flop"* will allow an input to LATCH in either of *two stable states* – the reset state represented by a 0 or the set stage represented by a 1. Through sequential digital logic, the flip-flop circuit may either go one or zero as two stable states. **ANSWER B**

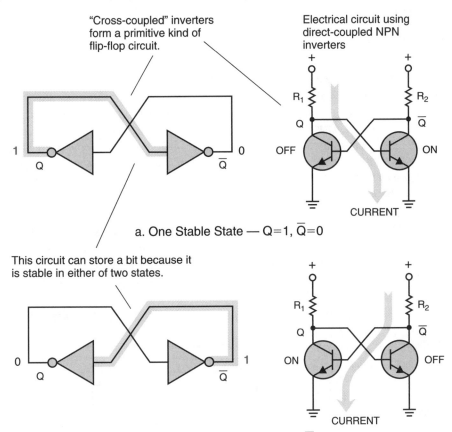

"Cross-coupled" inverters form a primitive kind of flip-flop circuit.

Electrical circuit using direct-coupled NPN inverters

a. One Stable State — Q=1, $\overline{Q}$=0

This circuit can store a bit because it is stable in either of two states.

b. Second Stable State — Q=0, $\overline{Q}$=1

Basic concept of the flip-flop, also called a bistable element or a static memory element.
Source: *Basic Electronics* © 1994, Master Publishing, Inc., Niles, Illinois

G7B02 Why do digital circuits use the binary number system?
A. Binary "ones" and "zeros" are easy to represent with an "on" or "off" state
B. The binary number system is most accurate
C. Binary numbers are more compatible with analog circuitry
D. All of these answers are correct

The flip-flop may either switch to an "on" state represented by a binary 1, or an "off" state, represented by a binary 0. ***Binary "ones" and "zeros" are easy to represent with an "on" or "off" state.*** **ANSWER A**

Elmer Point: Here is a very important point for your understanding of digital electronics: To write and keep track of the many possible combinations of high or low signals in digital information, the combinations are treated as binary numbers. Binary numbers are also called base two numbers. An example is shown below.

These days, most children learn about decimal numbers in elementary school. When written in ordinary decimal form, integers (whole numbers) are expressed as so many ones, so many tens, so many hundreds, and so forth. Each place in the number has a value ten times the place to the right. This requires using ten (for decimal) different symbols called decimal numerals or digits: 0, 1, 2, 3, 4, 5, 6, 7, 8, and 9. But in binary form, whole numbers are expressed as so many ones, so many twos, so many fours, and so forth. Each place in the number has twice the value of the next place to the right. Binary numbers use only two digits instead of ten: just 0 and 1.

So, in binary form, a number is written as a string of ones and zeroes. For instance, the integer one is written as 1. Two would be 10, which we read as "one-zero". It means a two and no one. Three would be 11 ("one-one", not eleven), meaning a two and a one. Four would be 100, called "one-zero-zero", meaning a four, no two, and no one. Five would be 101, meaning a four, no two, and a one. One hundred would be 1100100. That means a sixty-four, a thirty-two, no sixteen, no eight, a four, no two, and no one.

Each zero or one in a binary number is called a **binary digit**, *or bit for short. Bit also means a little piece of information. In fact, a bit is the smallest possible piece of information in a digital system. It expresses a choice between only two possibilities. In a binary number, for instance, a particular bit can say either, "Yes, there is a 64 in this number," or, "No, there is not a 64 in this number."*

In digital systems, binary numbers are used to write and keep track of the many possible combinations of the two electrical states.

Source: *Basic Electronics* © 1994, Master Publishing, Inc., Niles, Illinois

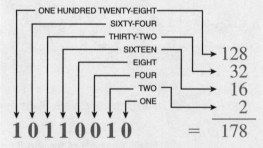

Binary Number:

Each place has a value **twice as great** as the next place to the right. **Two numerals or digits** are used:

0 (Zero = "No")
1 (One = "Yes")

These digits are called bits.

Binary System

G7B03 What is the output of a two-input NAND gate, given both inputs are "one"?

A. Two   C. Zero
B. One   D. Minus One

Ham operators designing flip-flop circuits will consult a "truth table" that allows them to view multiple flip-flop configurations. With the AND gate, both one's at the input will create a one on the output. However, it's just the opposite for a *NAND* gate – *two ones* on the input will lead to *zero on the output*. **ANSWER C**

The TTL Gate is a Positive-Logic NAND.

Function Table

A	B	Q
L	L	H
L	H	H
H	L	H
H	H	L

Positive Logic Truth Table

A	B	Q
0	0	1
0	1	1
1	0	1
1	1	0

The Output Q is Inverted from a Positive-Logic AND Gate

G7B04 What is the output of a NOR gate given that both inputs are "zero"?

A. Zero   C. Minus one
B. One   D. The opposite from the previous state

On the *NOR gate* with *both inputs zero*, the inverter will yield a *one on the output*. And it's just the opposite with an OR gate – two zeros on the input will yield a zero on the output. As you can see, with both these questions, the letter "N" in front of AND and OR means that the output will invert from the same input ones or zeros. There are only two questions regarding a NAND gate and a NOR gate, so gaze at these truth tables and go for the easy "opposite" answer. **ANSWER B**

G7B05 How many states are there in a 3-bit binary counter?

A. 3   C. 8
B. 6   D. 16

Each flip-flop requires two input pulses to generate one output pulse. Each time you add another flip-flop in series, you multiple the number of inputs required to generate one output by a factor of two. *2 x 2 x 2 = 8*. **ANSWER C**

G7B06 What is a shift register?

A. A clocked array of circuits that passes data in steps along the array
B. An array of operational amplifiers used for tri-state arithmetic operations
C. A digital mixer
D. An analog mixer

A *shift register passes flip-flop data* up or down to each other, rather than just an output. Parallel data may be converted to serial data, or the reverse, through a shift register. **ANSWER A**

G6C01 Which of the following is most often provided as an analog integrated circuit?
 A. NAND Gate
 B. Gallium Arsenide UHF Receiver "front end" Amplifier
 C. Frequency Counter
 D. Linear voltage regulator
Analog circuits are like sliding down a smooth banister at Grandpa's old house. Digital circuits would be like bumping down the stairs. *The linear voltage regulator* is found in an *analog integrated circuit* because of the absence of individual set levels. **ANSWER D**

G6C02 Which of the following is the most commonly used digital logic family of integrated circuits?
 A. RTL   C. CMOS
 B. TTL   D. PMOS
The *most common* digital logic family of integrated circuits is transistor-transistor logic, which may look OPEN as a high logic state. A logic high input in a *TTL* device operating with a 5 volt power supply is between 2.0 volts to 5.5 volts. **ANSWER B**

G6C03 Which of the following is an advantage of CMOS Logic integrated circuits compared to TTL logic circuits?
 A. Low power consumption
 B. High power handling capability
 C. Better suited for RF amplification
 D. Better suited for power supply regulation
The big advantage of CMOS Logic over TTL is that *CMOS* offers *low power consumption*. **ANSWER A**

G6C04 What is meant by the term ROM?
 A. Resistor Operated Memory   C. Random Operational Memory
 B. Read Only Memory   D. Resistant to Overload Memory
Your VHF/UHF handheld radio, and likely the modern high frequency radio you are ready to buy, contain *read only memory*. It is somewhat permanent, developed at the ham radio transceiver factory as band limits and tuning ranges. This memory does not need a battery backup. **ANSWER B**

G6C05 What is meant when memory is characterized as "non-volatile"?
 A. It is resistant to radiation damage
 B. It is resistant to high temperatures
 C. The stored information is maintained even if power is removed
 D. The stored information cannot be changed once written
Another type of memory is called RAM – Random Access Memory. In some older HF transceivers, a small button battery backs up all of those 100 channels of cool frequencies you have entered into random access memory. The internal battery needs to be changed every 5 years, because the memory is volatile and will disappear if there is no voltage to keep it alive. Non-volatile RAM is what we see in newer transceivers, and even though you *pull the plug*, and there is no internal battery backup, *non-volatile memory storage remains*. **ANSWER C**

Website Resources

▼ IF YOU'RE LOOKING FOR	▼ THEN VISIT
equipment reviews	www.hamoperator.com
Call sign lookups and more	www.qrz.com
ham radio gear, call sign look-ups	www.hamcall.net
QSO's and more	www.hamgallery.com
ham radio resources	www.dxzone.com
articles, reviews, etc.	www.eham.net
used equipment ads and more	www.qth.com
operating tips from Calgary hams	www.cara.ampr.org
directional antennas	www.arrowantennas.com
direction finding tips and more	www.homingin.com
free e-mail service for hams	www.qsl.net
ham radio for people with disabilities	www.handiham.org
Canada's amateur radio society	www.rac.ca
operating tips, ham news	www.hamquick.com
FCC amateur radio enforcement log	www.rainreport.com
California club with good tech articles	www.cvarc.org/faq.htm
antennas and related gear	www.natcommgroup.com
antennas, tuners, and such	www.sgcworld.com
every ham accessory known to man	www.mfjenterprises.com
CQ Magazine's website	www.cq-amateur-radio.com
radio propagation reports	www.dxworld.com/50prop.html
news & science reports about space	www.spacetoday.org
mobile antennas and accessories	www.hiqantennas.com
educational resources and links	www.ecjones.org
antennas, connectors, and more	www.cq73.com
linking ham radio via the internet	www.winlink.org
inductive components for RFI	www.amidoncorp.com
HF amplifiers, antennas, and more	www.ameritron.com
feedlines, connectors, & wire galore	www.cablexperts.com
copper ground strap, antenna masts	www.metal-cable.com
all kinds of antennas & accessories	www.antennaworld.com
antenna site with good technical data	www.cushcraft.com
excellent RF safety calculator	http://n5xu.ae.utexas.edu/rfsafety
technical resources for hams	www.csvhfs.org

Elmer Point: *Your Granddad ham likely knew that a chap named Georg Simon Ohm (1789-1854) experimented with electricity and discovered that the resistance (R) of a conductor deptends on its length in feet, cross-sectional area in circular mils, and the resistivity, which is a parameter that depends on the molecular structure of the conductor and its temperature. Ohm's Law states:*

The current in an electrical circuit is directly proportional to the voltage and inversely proportional to the resistance.

The Ohm's Law and Power Circle shown here includes 12 equations that allow us to solve for voltage,(E), current (I), resistance (R), and power (P) if we know the other values. There isn't much tough math in this section, but keep this page bookmarked to help you solve common electrical formulas.

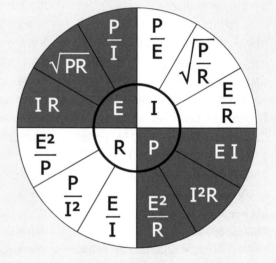

To solve for Voltage (E):

$E = I$ (current) $\times R$ (resistance)

$E = \sqrt{P \text{ (power)} \times R \text{ (resistance)}}$

$E = P$ (power) $\div I$ (current)

To solve for Current (I):

$I = P$ (power) $\div E$ (voltage)

$I = \sqrt{P \text{ (power)} \div R \text{ (resistance)}}$

$I = E$ (voltage) $\div R$ (resistance)

To solve for Resistance (R):

$R = E^2$ (voltage squared) $\div P$ (power)

$R = P$ (power) $\div I^2$ (current squared)

$R = E$ (voltage) $\div I$ (current)

To solve for Power (P):

$P = E$ (voltage) $\times I$ (current)

$P = I^2$ (current squared) $\times R$ (resistance)

$P = E^2$ (voltage squared) $\div R$ (resistance)

Source: The ARRL Handbook, 2207, © 2006, American Radio Relay League

G5B12 What would be the voltage across a 50-ohm dummy load dissipating 1200 watts?

A. 173 volts   C. 346 volts
B. 245 volts   D. 692 volts

When you work with high-frequency mobile antennas or, in this question, a high-frequency 50-ohm dummy load, you may wish to calculate the voltage across the load to make sure things won't arc over or to determine the maximum value of voltage which can be applied across the load without exceeding its power rating. Here is your formula:

$$E = \text{Square root of P x R} = \sqrt{P \times R}$$
$$E = \text{Square root of } 50 \times 1200 = \sqrt{50 \times 1200}$$
$$E = \text{Square root of } 60,000 = \sqrt{60,000}$$
$$E = 244.948$$

This is an easy one to work out on a calculator – and yes, calculators are allowed in the exam room. Here are the keystrokes: Clear, Clear, 50 x 1200 = 60,000. Now simply tap the square root sign and, voila, the correct answer pops out at (approximately) 245 volts, which shows up as Answer B on your exam. Don't forget, examiners are allowed to scramble the A, B, C, D order, so simply look for *245 volts* as the correct answer. **ANSWER B**

G5B03 How many watts of electrical power are used if 400 VDC is supplied to an 800-ohm load?

A. 0.5 watts   C. 400 watts
B. 200 watts   D. 3200 watts

The equation for this problem is $P = E^2 \div R$. 400^2 is 160,000. That divided by R (800) is 200, so the answer is *200 watts*. Calculator keystrokes are: Clear, 400 x 400 ÷ 800 = 200. **ANSWER B**

G5B04 How many watts of electrical power are used by a 12-VDC light bulb that draws 0.2 amperes?

A. 2.4 watts   C. 6 watts
B. 24 watts   D. 60 watts

Easy equation – power is equal to voltage times current ($P = E \times I$). Multiply volts times amps, and you end up with 2.4 watts. Using the magic circle for power, you see that power is equal to E x I. Calculator keystrokes are: Clear, *12 x 0.2 = 2.4 (in watts)*. **ANSWER A**

G5B05 How many watts are being dissipated when a current of 7.0 milliamperes flows through 1.25 kilohms?

A. Approximately 61 milliwatts   C. Approximately 11 milliwatts
B. Approximately 39 milliwatts   D. Approximately 9 milliwatts

Use the *current and resistance version of the magic power circle* for the equation $P = I^2 \times R$. $I = 0.007$ amperes and $R = 1250$ ohms. Your answer will come out 0.061 watts. This is converted to 61 milliwatts by moving the decimal point 3 places to the right. Calculator keystrokes are: Clear, *0.007 x 0.007 x 1250 = 0.06125 (in watts)*. **ANSWER A**

G5B07 Which measurement of an AC signal is equivalent to a DC voltage of the same value?

A. The peak-to-peak value C. The RMS value
B. The peak value D. The reciprocal of the RMS value

Root Mean Square (RMS) measurement of an AC signal *(also called the effective value of an AC voltage)* is the same as a DC voltage of the same value.
ANSWER C

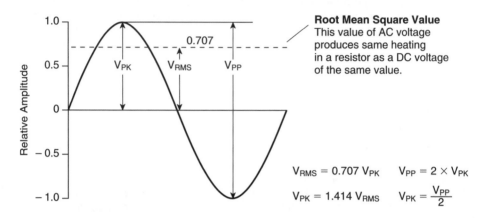

Root Mean Square Value
This value of AC voltage produces same heating in a resistor as a DC voltage of the same value.

$V_{RMS} = 0.707\ V_{PK}$ $V_{PP} = 2 \times V_{PK}$

$V_{PK} = 1.414\ V_{RMS}$ $V_{PK} = \dfrac{V_{PP}}{2}$

RMS (V$_{RMS}$), Peak, (V$_{PK}$), and Peak-to-Peak (V$_{··}$) Voltage

G5B08 What is the peak-to-peak voltage of a sine wave that has an RMS voltage of 120 volts?

A. 84.8 volts C. 240.0 volts
B. 169.7 volts D. 339.4 volts

120 volts multiplied by 1.41 (average to peak) with the result multiplied by 2 (peak to peak) results in *339.4 volts*, which is what you might see on an oscilloscope sampling voltage out of your wall socket. You didn't realize your house wiring was that good, huh? **ANSWER D**

G5B09 What is the RMS voltage of sine wave with a value of 17 volts peak?

A. 8.5 volts C. 24 volts
B. 12 volts D. 34 volts

To go from 17 volts peak down to average voltage, multiply by 0.707. You end up with *12 volts AC*. **ANSWER B**

G7A15 Which of the following is an advantage of a switched-mode power supply as compared to a linear power supply?

A. Faster switching time makes higher output voltage possible
B. Fewer circuit components are required
C. High frequency operation allows the use of smaller components
D. All of these choices are correct

The switching power supply is about the same price as the old, big, heavy transformer power supplies for 20 amps or 35 amps, 12-volt DC output. The switching power supply uses a relatively *high-frequency oscillator* at a frequency where small, lightweight and low-cost *miniature transformers* create a relatively smooth DC power output. But one concern of the switching power supply is the proximity of your new ham high-frequency antenna system. Your antenna must be at least 10 feet or further away from the switching power supply so as to not pick up broad-band noise from the switching power supply. **ANSWER C**

Modern switching power supplies incorporate "crowbar protection" to provide overvoltage protection

---

G7A14 Which of the following is a desirable characteristic for capacitors used to filter the DC output of a switching power supply?
A. Low equivalent series resistance
B. High equivalent series resistance
C. Low Temperature coefficient
D. High Temperature coefficient

Switching power supplies are those lightweight wonders that don't require a big, heavy transformer on the inside. Unlike the big transformer power supplies that use large electrolytic capacitors, switching power supplies employ *smaller capacitors WITH LOW EQUIVALENT SERIES RESISTANCE* that make them more compatible with the frequency wave form switching techniques inside the switching supply. **ANSWER A**

G6B04 When two or more diodes are connected in parallel to increase current handling capacity, what is the purpose of the resistor connected in series with each diode?
A. The resistors ensure the thermal stability of the power supply
B. The resistors regulate the power supply output voltage
C. The resistors ensure that one diode doesn't carry most of the current
D. The resistors act as swamping resistors in the circuit

Adding a *resistor* in series with two or more diodes connected in PARALLEL will help *balance*, equally, the *amount of current that each diode will pass*.
ANSWER C

G6A11 What is the common name for a capacitor connected across a transformer secondary that is used to absorb transient voltage spikes?
A. Clipper capacitor   C. Feedback capacitor
B. Trimmer capacitor   D. Suppressor capacitor

If your new power supply suddenly fails after an obvious power surge at your home, check the *suppressor capacitors* for a short. These suppressor capacitors *filter transient voltage spikes* across the transformer's secondary winding. **ANSWER D**

G7A16 What portion of the AC cycle is converted to DC by a half-wave rectifier?

A. 90 degrees   C. 270 degrees
B. 180 degrees   D. 360 degrees

The *half-wave* rectifier uses only half of the cycle, which is *180 degrees*.
ANSWER B

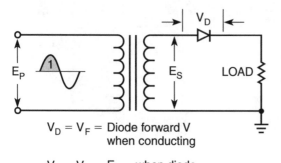

$V_D = V_F =$ Diode forward V
when conducting

$V_D = V_R = E_{SPK}$ when diode
is not conducting

Conducts only on positive cycle.
No conduction on negative cycle.

Reverse voltage across diode
is E_{SPK}, the peak voltage
of the secondary voltage.

Half-Wave Rectifier

G7A17 What portion of the AC cycle is converted to DC by a full-wave rectifier?

A. 90 degrees   C. 270 degrees
B. 180 degrees   D. 360 degrees

A *full-wave* rectifier is much more efficient because it uses all *360 degrees*. A full-wave rectifier output also is much easier to filter to provide pure DC voltage. (See page 122) **ANSWER D**

G7A18 What is the output waveform of an unfiltered full-wave rectifier connected to a resistive load?

A. A series of DC pulses at twice the frequency of the AC input
B. A series of DC pulses at the same frequency as the AC input
C. A sine wave at half the frequency of the AC input
D. A steady DC voltage

A *full-wave* rectifier gives a much smoother *pulsating DC* to filter than a half-wave rectifier because the full-wave rectified half sine waves are *double* at the line frequency. **ANSWER A**

G6B01 What is the peak-inverse-voltage rating of a rectifier?

A. The maximum voltage the rectifier will handle in the conducting direction
B. 1.4 times the AC frequency
C. The maximum voltage the rectifier will handle in the non-conducting direction
D. 2.8 times the AC frequency

The peak-inverse-voltage *(PIV)* rating of a power supply rectifier is the *maximum voltage* it will handle in the *non-conducting direction*. **ANSWER C**

G7A03 What should be the minimum peak-inverse-voltage rating of the rectifier in a full-wave power supply?

A. One-quarter the normal output voltage of the power supply
B. Half the normal output voltage of the power supply
C. Double the normal peak output voltage of the power supply
D. Equal to the normal output voltage of the power supply

Double or nothing! The *peak-inverse-voltage rating* of *rectifier diodes* normally is *double for a full-wave power supply*. This is because the diode is seeing the entire secondary winding, as opposed to just half of the secondary winding in the opposite direction. It's always a good idea to double the voltage rating of any type of capacitor or diode going into a power supply circuit. **ANSWER C**

G7A04 What should be the approximate minimum peak-inverse-voltage rating of the rectifier in a half-wave power supply?

A. One-half the normal peak output voltage of the power supply
B. Half the normal output voltage of the power supply
C. Equal to the normal output voltage of the power supply
D. Two times the normal peak output voltage of the power supply

At least one time, but to be much more reliable, *two times the peak voltage output.* **ANSWER D**

G6B02 What are the two major ratings that must not be exceeded for silicon-diode rectifiers?

A. Peak inverse voltage; average forward current
B. Average power; average voltage
C. Capacitive reactance; avalanche voltage
D. Peak load impedance; peak voltage

Be sure to *never exceed the peak-inverse-voltage rating and the average forward current* rating of silicon-diode rectifiers used in power supplies. **ANSWER A**

G7A02 What components are used in a power-supply filter network?

A. Diodes
B. Transformers and transistors
C. Quartz crystals
D. Capacitors and inductors

In power supplies, transformers supply the voltage and current, diodes rectify, and *capacitors and inductors filter.* A bleeder resistor protects. **ANSWER D**

G6A04 Which of the following is an advantage of an electrolytic capacitor?

A. Tight tolerance
B. Non-polarized
C. High capacitance for given volume
D. Inexpensive RF capacitor

We usually find the electrolytic capacitor in the power supply section of our new rig. The electrolytic capacitor is polarized with a positive and negative connection point. The electrolytic offers *high capacitance for given volume.* (Volume = size) **ANSWER C**

G6A02 What type of capacitor is often used in power-supply circuits to filter the rectified AC?

A. Disc ceramic
B. Vacuum variable
C. Mica
D. Electrolytic

Rectified AC is a form of pulsating DC. *Electrolytic capacitors* usually are used as *filters* to smooth pulsating DC because they offer large amounts of capacity in a small size. The big problem with electrolytic capacitors – especially old ones – is that they dry out and lose their capacitance. **ANSWER D**

G6A12 What is the common name for an inductor used to help smooth the DC output from the rectifier in a conventional power supply?

A. Back EMF choke   C. Charging inductor

B. Repulsion coil   D. Filter choke

The half wave rectifier will produce a small amount of AC ripple, identical to the line input frequency, normally 60 Hz. Full wave rectifiers will produce twice the 60 Hz ripple, 120 Hz. The *filter choke inductor* is much more efficient at this double AC frequency to smooth out the AC ripple imposed on the DC line. **ANSWER D**

G7A01 What safety feature does a power-supply bleeder resistor provide?

A. It acts as a fuse for excess voltage

B. It discharges the filter capacitors

C. It removes shock hazards from the induction coils

D. It eliminates ground-loop current

When you turn off your big rig, it dims down and then cycles off completely. This slow decay of voltage is from the filter capacitors that are slowly being discharged by the *bleeder resistors*. This is a *safety feature*, and also helps provide voltage regulation. **ANSWER B**

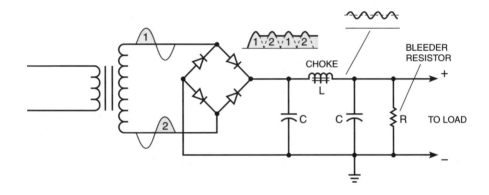

Full-Wave Power Supply with Bleeder Resistor

G4B07 What is an advantage of a digital voltmeter as compared to an analog voltmeter?

A. Better for measuring computer circuits

B. Better for RF measurements

C. Significantly better precision for most uses

D. Faster response

Your buddy discovers you are a brand new General and gives you several gel cell 12 volt batteries. To see which one has the absolute best charge, use a digital volt meter to get a *more precise digital readout* of the battery voltage. A neat little analog volt meter might not show you subtle changes in battery terminal voltage. **ANSWER C**

G4B16 Why is high input impedance desirable for a voltmeter?
A. It improves the frequency response
B. It decreases battery consumption in the meter
C. It improves the resolution of the readings
D. It decreases the loading on circuits being measured

New digital volt meters have a relatively high *input impedance*, so as *not to load down the circuit being measured*. If you are measuring a fraction of a volt, you don't want to use an older volt meter that might significantly load down the circuit and give you a bogus reading. **ANSWER D**

G4E07 When might a lead-acid storage battery give off explosive hydrogen gas?
A. When stored for long periods of time
B. When being discharged
C. When being charged
D. When not placed on a level surface

I run my home station off of a lead acid storage battery to keep me on the air in case of a blackout. The battery is outside because it gives off *explosive hydrogen gas when it is being charged* by solar panels, and it works very well. Never use a storage battery inside your ham shack because of the danger of explosive hydrogen gas. **ANSWER C**

G6B14 What is the minimum allowable discharge voltage for maximum life of a standard 12 volt lead acid battery?
A. 6 volts
B. 8.5 volts
C. 10.5 volts
D. 12 volts

When operating your equipment with a 12 volt lead acid automobile battery during Field Day, make sure you never pull your battery below *10.5 volts*. If you do, you will shorten the maximum life of that battery, and any transmitter running at 10.5 volts (instead of 12 volts) will usually sound distorted on high frequency. **ANSWER C**

G6B16 Which of the following is a rechargeable battery?
A. Carbon-zinc
B. Silver oxide
C. Nickel Metal Hydride
D. Mercury

A rechargeable battery is the nickel metal hydride chemistry. Nickel cadmium may also be recharged but these cells are slowly fading from the marketplace because of the problem of proper disposal of Cadmium. The *nickel metal hydride* battery pack offers almost twice the capacity as an older nickel cadmium pack. **ANSWER C**

G6B15 When is it acceptable to recharge a carbon-zinc primary cell?
A. As long as the voltage has not been allowed to drop below 1.0 volt
B. When the cell is kept warm during the recharging period
C. When a constant current charger is used
D. Never

Disposable flashlight "D" cells are never intended for recharge. Attempts to recharge carbon zinc primary cells or newer alkaline batteries may lead to the battery venting dangerous gas. Unless the battery specifically states "rechargeable," *do NOT try to recharge NON-rechargeable cells*. **ANSWER D**

G6B13 What is an advantage of the low internal resistance of Nickel Cadmium batteries?
A. Long life
C. High voltage
B. High discharge current
D. Rapid recharge

The nickel cadmium battery is a low-cost, rechargeable voltage source for handheld radios as well as for some QRP (low power output) high frequency transceivers. The *low internal resistance* of the nickel cadmium battery allows for *high discharge current* when transmitting. The disadvantage of low internal resistance is the slight self-discharge between uses of your radio equipment. Before going out on Field Day, be sure to cycle your nickel cadmium battery pack several times, and end the cycle with a good charge. **ANSWER B**

G4E08 What is the name of the process by which sunlight is changed directly into electricity?
A. Photovoltaic conversion
C. Photosynthesis
B. Photon emission
D. Photon decomposition

You can change sunlight into VOLTAGE, and notice the word "volt" in the term *photovoltaic conversion.* **ANSWER A**

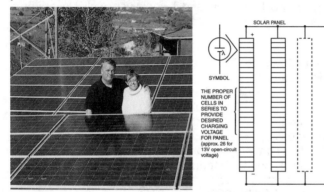

Solar Panel array for
Charging Storage Batteries

Schematic of Solar Panel
for Charging Storage Battery

G4E10 Which of these materials is used as the active element of a solar cell?
A. Doped Silicon
C. Doped Platinum
B. Nickel Hydride
D. Aluminum nitride

Solar and silicone – what a great combination! Solar cells use *doped silicon* as the active element. **ANSWER A**

G4E09 What is the approximate open-circuit voltage from a modern, well illuminated photovoltaic cell?
A. 0.02 VDC
C. 0.2 VDC
B. 0.5 VDC
D. 1.38 VDC

A complete photovoltaic cell will yield *0.5 volts* direct current. A solar panel is made up of a series-parallel connection of these cells in order to charge your storage battery system. **ANSWER B**

G4E11 Which of the following is a disadvantage to using wind power as the primary source of power for an emergency station?

A. The conversion efficiency from mechanical energy to electrical energy is less that 2 percent

B. The voltage and current ratings of such systems are not compatible with amateur equipment

C. A large energy storage system is needed to supply power when the wind is not blowing

D. All of these choices are correct

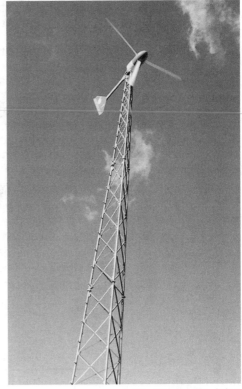

The wind doesn't blow all the time, so wind power is not a good primary source for your emergency communications station. When the wind isn't blowing, you need a *huge bank of batteries to keep your station on the air*.
ANSWER C

No Wind – No Juice

G4E13 Why would it be unwise to power your station by back feeding the output of a gasoline generator into your house wiring by connecting the generator through an AC wall outlet?

A. It might present a hazard for electric company workers

B. It is prone to RF interference

C. It may disconnect your RF ground

D. None of the above; this is an excellent expedient

Never, ever try to feed 110 VAC back into your home during a power outage using that brand new gas-powered generator out in the RV. Even though you may have shut off all of your circuit breakers, it is always a *bad idea to backwards feed electricity into your house – someone could get hurt*. There are special isolators and special controllers available for those hams who may wish to supplement commercial power in their house with auxiliary power from wind turbine and solar arrays. **ANSWER A**

G4E03 Which of the following power connections would be the best for a 100-watt HF mobile installation?
A. A direct, fused connection to the battery using heavy gauge wire
B. A direct, fused connection to the alternator or generator using heavy gauge wire
C. A direct, fused connection to the battery using resistor wire
D. A direct, fused connection to the alternator or generator using resistor wire

As a General Class operator, you will probably be running a 100-watt, high-frequency, mobile transceiver in your car. You cannot rely on the car's 12-volt accessory or "power point" socket wiring to support the necessary 20-amp (minimum) power demands from your new radio. *Wire your red and black power leads directly to the battery* using heavy-gauge wire. Fuse both the red and the black power leads close to the battery terminals. **ANSWER A**

Elmer Point: *To go mobile with your 100-watt HF ham transceiver, you'll need to run the red and black power wires directly to the positive and negative terminals on your car or truck battery. You should have separate fuses right next to the battery terminal connections for safety. While automotive sound systems sometimes use the vehicle chassis as the negative black wire return, commercial two-way radio installers always say it's best to run directly to the positive and negative terminals on the battery. The chassis of your HF radio also should be well grounded to your vehicle frame.*

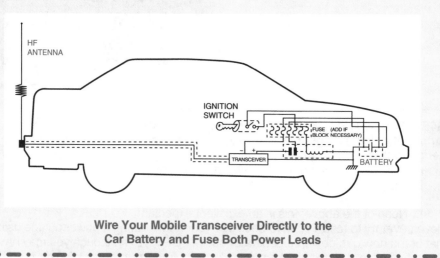

Wire Your Mobile Transceiver Directly to the
Car Battery and Fuse Both Power Leads

G4E04 Why is it best NOT to draw the DC power for a 100-watt HF transceiver from an automobile's cigarette lighter socket?
A. The socket is not wired with an RF-shielded power cable
B. The socket's wiring may be inadequate for the current being drawn by the transceiver
C. The DC polarity of the socket is reversed from the polarity of modern HF transceivers
D. The power from the socket is never adequately filtered for HF transceiver operation

While you might be tempted to grab 12 volts from an automobile cigarette lighter socket, DON'T. Sure, the radio will work for a few minutes, but after a little bit of transmitting *the cigarette lighter receptacle wiring will get red hot from over-drawing current*, and quite possibly ignite more than what it was supposed to light up! **ANSWER B**

G4D11 What is the main reason to use keyed connectors over non-keyed types?
  A. Prevention of use by unauthorized persons
  B. Reduced chance of damage due to incorrect mating
  C. Higher current carrying capacity
  D. All of these choices are correct

Metal microphone plugs and jacks all have a protruding ridge or a channel so that the metal microphone plug fits properly into the receptacle. They call this a *"keyed" connector* and it will *reduce the chance of accidentally bending fragile pins* when you go to plug in your microphone connector. **ANSWER B**

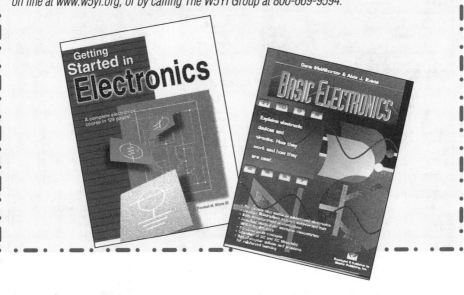

Elmer Point: *Want to learn more about electricity and how electronics work? Here are two books I recommend highly for your self-education!*

Getting Started in Electronics *by Forrest M. Mims III is a true classic. It is used by a wide range of people – from junior-high teachers to the U.S. Army to teach the fundamentals of electricity and electronics.*

Basic Electronics *by Alvis J. Evans and Gene McWhorter goes a little deeper into the topic, and includes end of chapter quizzes and worked-out problems to teach you in detail the various aspects of electronics.*

Either book – or both – will give you a solid grounding in the theory, science, and practical applications of electronics. You can get you copies of the books at your local ham radio store, on line at www.w5yi.org, or by calling The W5YI Group at 800-669-9594.

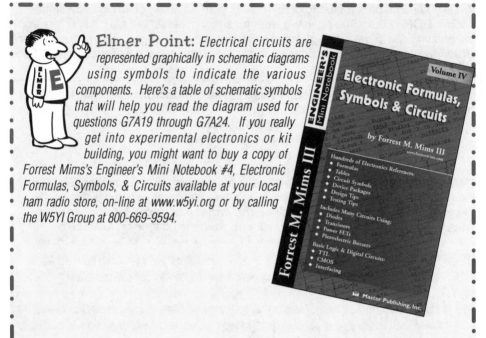

Elmer Point: Electrical circuits are represented graphically in schematic diagrams using symbols to indicate the various components. Here's a table of schematic symbols that will help you read the diagram used for questions G7A19 through G7A24. If you really get into experimental electronics or kit building, you might want to buy a copy of Forrest Mims's Engineer's Mini Notebook #4, Electronic Formulas, Symbols, & Circuits available at your local ham radio store, on-line at www.w5yi.org or by calling the W5YI Group at 800-669-9594.

SCHEMATIC SYMBOLS

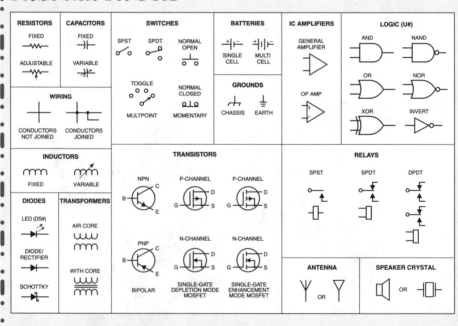

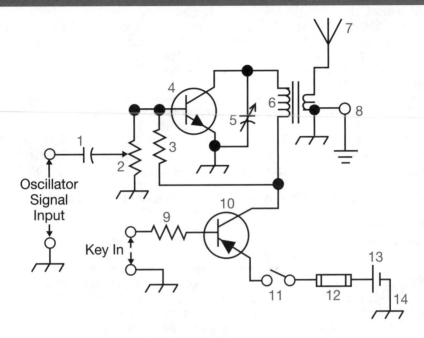

Graphic G7-1 for 2007-11 General Class Question Pool

G7A19 Which symbol in figure G7-1 represents a fixed resistor?
    A. Symbol 2   C. Symbol 3
    B. Symbol 6   D. Symbol 12

In graphic G7-1, we can spot two *fixed value resistors, symbol 3* and symbol 9, with symbol 3 as the correct answer. We call this a schematic symbol, part of the overall schematic diagram. **ANSWER C**

G7A20 Which symbol in figure G7-1 represents a single cell battery?
    A. Symbol 5   C. Symbol 8
    B. Symbol 12   D. Symbol 13

Let's now study the schematic diagram and look for the schematic symbol of a single cell battery. Hint: bottom right. Notice the smaller straight vertical line as the battery negative terminal, going to chassis ground (symbol 14). The larger vertical line is the battery positive, going to the fuse (symbol 12). So *symbol 13 is the single cell battery* in the schematic diagram. **ANSWER D**

G7A21 Which symbol in figure G7-1 represents a NPN transistor?
    A. Symbol 2   C. Symbol 10
    B. Symbol 4   D. Symbol 12

We have a pair of transistors in this diagram, and let's zero in on the top transistor. This is an NPN bipolar transistor, and notice that the emitter arrow is NOT POINTING IN. This is how you can distinguish this transistor as an NPN (not pointing in) versus the lower transistor, a PNP, where the arrow is indeed pointing in. So, *symbol 4 is the NPN transistor*. **ANSWER B**

G7A22 Which symbol in figure G7-1 represents a variable capacitor?
  A. Symbol 2   C. Symbol 5
  B. Symbol 11   D. Symbol 12

Our search of this schematic diagram is for a variable capacitor and you can spot the capacitor with its concave curve and straight line. But notice the arrow? This means *symbol 5 is indeed a variable capacitor*. Just like symbol 2, with an arrow as a variable resistor. But they asked for the variable capacitor, so go for symbol 5. **ANSWER C**

G7A23 Which symbol in figure G7-1 represents a transformer?
  A. Symbol 6   C. Symbol 10
  B. Symbol 4   D. Symbol 2

Now we look for the classic iron core *transformer*, and there it is at *symbol 6*. The two vertical lines indicate a core between the primary and secondary windings. **ANSWER A**

G7A24 Which symbol in figure G7-1 represents a single pole switch?
  A. Symbol 2   C. Symbol 11
  B. Symbol 3   D. Symbol 12

The single pole switch looks like a *switch at symbol 11*. More technically, it is a single pole single throw switch. **ANSWER C**

Elmer Point: *Here is a set of questions that asks about the total value of resistors, capacitors, and inductors connected in SERIES and in PARALLEL. Resistors (R) and Inductors (L) act the same way. Capacitors (C) act in the opposite way. Here are the formulas to calculate the total value of these components:*

$$\text{Resistors in SERIES simply add up: } R_{total} = R_1 + R_2 + R_3$$

$$\text{Inductors in SERIES simply add up: } L_{total} = L_1 + L_2 + L_3$$

Resistors and Inductors in PARALLEL combine with a resulting total value that is always LESS than the value of the lowest value resister in parallel, for easy problem solving! Here are the formulas:

$$\text{when there are 2 resistors or inductors: } R_{total} = \frac{R_1 \times R_2}{R_1 + R_2} \quad \text{or} \quad L_{total} = \frac{L_1 \times L_2}{L_1 + L_2}$$

$$\text{when there are 3 or more resistors: } R_{total} = \frac{1}{\dfrac{1}{R_1} + \dfrac{1}{R_2} + \dfrac{1}{R_3}} \quad \text{or} \quad L_{total} = \frac{1}{\dfrac{1}{L_1} + \dfrac{1}{L_2} + \dfrac{1}{L_3}}$$

Capacitors in PARALLEL simply add up: $C_{total} = C_1 + C_2 + C_3$

Capacitors in SERIES combine with a resulting total value that is always LESS than the value of the lowest value capacitor in series, for easy problem solving! Here is the formula:

$$\text{when there are 2 capacitors: } C_{total} = \frac{C_1 \times C_2}{C_1 + C_2}$$

$$\text{when there are 3 or more capacitors: } C_{total} = \frac{1}{\dfrac{1}{C_1} + \dfrac{1}{C_2} + \dfrac{1}{C_3}}$$

G5C04 What is the total resistance of three 100-ohm resistors in parallel?
A. 0.30 ohms   C. 33.3 ohms
B. 0.33 ohms   D. 300 ohms

If we have three equal value resistors in parallel, we will have three individual paths for current to flow, decreasing the circuit's ohmic resistance by 2/3. You can do this one in your head – *1/3 the resistance of each 100 ohm resistor is 33.3 ohms*. **ANSWER C.**
If you want to do it the long way, here is the formula:

$$R_T = \cfrac{1}{\left(\cfrac{1}{R_1} + \cfrac{1}{R_2} + \cfrac{1}{R_3}\right)}$$

G6A01 What will happen to the resistance if the temperature of a carbon resistor is increased?
A. It will increase by 20% for every 10 degrees centigrade
B. It will stay the same
C. It will change depending on the resistor's temperature coefficient rating
D. It will become time dependent

Heating a carbon resistor always decreases its resistance. However, the amount of change in resistance for any particular temperature *change depends on its temperature coefficient*, which depends on the materials used in the resistor's construction. **ANSWER C**

G6A13 What type of component is a thermistor?
A. A resistor that is resistant to changes in value with temperature variations
B. A device having a controlled change in resistance with temperature variations
C. A special type of transistor for use at very cold temperatures
D. A capacitor that changes value with temperature

When you earn your new General license, you may hook up with another station thousands of miles away who will ask you what the weather is. Likely you have one of those new wireless weather stations, and the temperature sensor uses a component called a THERMISTOR. The *thermistor* is a resistor which is designed to MAXIMIZE *changes in value with temperature variations*. This makes it a great reference for slight changes in temperature. **ANSWER B**

G5C10 What is the inductance of three 10 millihenry inductors connected in parallel?
A. 0.30 Henrys   C. 3.3 millihenrys
B. 3.3 Henrys   D. 30 millihenrys

Treat inductors like resistors when working either series or parallel problems. This one you can do in your head. *Three 10 millihenry inductors in parallel*, total inductance will be *1/3 or 3.3* millihenrys when connected in parallel. **ANSWER C**

G5B02 How does the total current relate to the individual currents in each branch of a parallel circuit?
A. It equals the average of each branch current
B. It decreases as more parallel branches are added to the circuit
C. It equals the sum of the currents through each branch
D. It is the sum of the reciprocal of each individual voltage drop

If you *add up the current in each branch*, you will come up with the *total current in a parallel circuit*. **ANSWER C**

G5C05 What is the value of each resistor if three equal value resistors in parallel produce 50 ohms of resistance, and the same three resistors in series produce 450 ohms?

A. 1500 ohms
C. 150 ohms
B. 90 ohms
D. 175 ohms

If we have three like resistors in parallel that together produce 50 ohms of resistance, each individual resistor would have a value of *150 ohms (50+50+50)*. Now check your answer – if we have three 150 ohm resistors in series, the total resistance adds up to 450 ohms. **ANSWER C**

G5C15 What is the total resistance of a 10 ohm, a 20 ohm, and a 50 ohm resistor in parallel?

A. 5.9 ohms
C. 10000 ohms
B. 0.17 ohms
D. 80 ohms

While you can solve this problem with a big formula, always remember that for resistors with differing values in parallel, just like capacitors with differing values in series, the resulting answer will always be less than the smallest value component. *5.9 ohms* is a logical answer to solve for when you work the big long formula all the way out. **ANSWER A.** Use the same formula shown at G5C04 on page 131.

G5C16 What component should be added to an existing resistor in a circuit to increase circuit resistance?

A. A resistor in parallel
C. A capacitor in series
B. A resistor in series
D. A capacitor in parallel

Another easy one here – to add *more resistance* to a circuit, we *add resistor(s) in series*. **ANSWER B**

G5C08 What is the equivalent capacitance of two 5000 picofarad capacitors and one 750 picofarad capacitor connected in parallel?

A. 576.9 picofarads
C. 3583 picofarads
B. 1733 picofarads
D. 10750 picofarads

Calculating *total capacitance in parallel* is easy – it *is the sum of each individual capacitor*. 5000 + 5000 (remember they say two) + 750 = 10750 picofarads. CAP = capacitors in parallel simply add up. The formula is:
$C_T = C_1 + C_2 + C_3$. **ANSWER D**

G5C09 What is the capacitance of three 100 microfarad capacitors connected in series?

A. 0.30 microfarads
C. 33.3 microfarads
B. 0.33 microfarads
D. 300 microfarads

We have three like capacitors in series – since they are in series, like resistors in parallel, *total capacitance will be 1/3 of each 100 microfarad cap. 33.3 microfarads* is an answer you can do in your head! **ANSWER C.** The formula is:

$$C_T = \cfrac{1}{\left(\cfrac{1}{C_1} + \cfrac{1}{C_2} + \cfrac{1}{C_3}\right)}$$

G5C11 What is the inductance of a 20 millihenry inductor in series with a 50 millihenry inductor?

A. 0.07 millihenrys
C. 70 millihenrys
B. 14.3 millihenrys
D. 1000 millihenrys

Just like resistors add up in series, so do inductors. *20 + 50 = 70 millihenrys.* **ANSWER C**

G5C12 What is the capacitance of a 20 microfarad capacitor in series with a 50 microfarad capacitor?

A. 0.07 microfarads   C. 70 microfarads
B. 14.3 microfarads   D. 1000 microfarads

Calculating for total capacitance in series is much like the same formula for calculating unlike resistors in parallel: $(C_1 \times C_2) \div (C_1 + C_2)$. In this equation, the total capacitance will always be less than the smaller capacitor, so *14.3 microfarads* can be confirmed as the correct answer. **ANSWER B**

G5C13 What component should be added to a capacitor in a circuit to increase the circuit capacitance?

A. An inductor in series   C. A capacitor in parallel
B. A resistor in series   D. A capacitor in series

If we need to add some additional capacitance to a circuit which already has a fixed capacitor, we could *add a second capacitor in PARALLEL.* **ANSWER C**

G5C14 What component should be added to an inductor in a circuit to increase the circuit inductance?

A. A capacitor in series   C. An inductor in parallel
B. A resistor in parallel   D. An inductor in series

If we need to increase the inductance (L) of a circuit, we would simply add *another inductor in SERIES.* **ANSWER D**

G5A02 What is reactance?

A. Opposition to the flow of direct current caused by resistance
B. Opposition to the flow of alternating current caused by capacitance or inductance
C. A property of ideal resistors in AC circuits
D. A large spark produced at switch contacts when an inductor is de-energized

Inductive reactance is the opposition to AC caused by inductors. Capacitive reactance is the opposition to AC caused by capacitors. Both reactances vary with frequency. When there is an inductor and a capacitor in the same circuit, there is a special frequency, called the resonant frequency, where the inductive reactance equals the capacitive reactance. **ANSWER B**

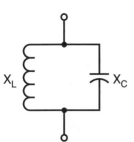

$$X_L = 2\pi f L$$

$$X_C = \frac{1}{2\pi f C}$$

The resonant frequency of a circuit is:

$$f_r = \frac{1}{2\pi \sqrt{LC}}$$

The resonant frequency is the frequency where $X_L = X_C$.

$$\therefore 2\pi f L = \frac{1}{2\pi f C}$$

$$f^2 = \frac{1}{(2\pi L)(2\pi C)}$$

$$f^2 = \frac{1}{(2\pi)^2 LC}$$

$$\therefore f_r = \frac{1}{2\pi \sqrt{LC}}$$

Resonant Frequency

G5A03 Which of the following causes opposition to the flow of alternating current in an inductor?
A. Conductance
B. Reluctance
C. Admittance
D. Reactance

Think of an inductor as a coil of wire. Its opposition to AC is called *inductive reactance*, identified as X_L in ohms. $X_L = 2\pi fL$ where f is the frequency in **hertz,** L is the inductance in **henries,** and π is 3.14. X_L increases as frequency increases. **ANSWER D**

G5A09 What unit is used to measure reactance?
A. Farad
B. Ohm
C. Ampere
D. Siemens

The *ohm* is the unit of measurement for reactance as well as resistance. **ANSWER B**

G5A04 Which of the following causes opposition to the flow of alternating current in a capacitor?
A. Conductance
B. Reluctance
C. Reactance
D. Admittance

A capacitor has plates separated by an insulating dielectric. Its opposition to AC is called *capacitive reactance*, identified as X_C in ohms. $X_C = 1 \div 2\pi fC$ where f is the frequency in **hertz,** C is the capacitance in **farads,** and π is 3.14. X_C decreases as frequency increases. **ANSWER C**

G5A06 How does a capacitor react to AC?
A. As the frequency of the applied AC increases, the reactance decreases
B. As the frequency of the applied AC increases, the reactance increases
C. As the amplitude of the applied AC increases, the reactance increases
D. As the amplitude of the applied AC increases, the reactance decreases

Capacitors offer reactance to AC inversely proportional to the frequency. Capacitors have high reactance at low frequencies and low reactance at high frequencies. Remember, *as f increases, X_C decreases* ($X_C = 1 \div 2\pi fC$). **ANSWER A**

FIXED   VARIABLE   ELECTROLYTIC

a. Symbols

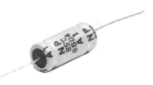

b. Physical Parts

Capacitors

G5A05 How does a coil react to AC?
A. As the frequency of the applied AC increases, the reactance decreases
B. As the amplitude of the applied AC increases, the reactance increases
C. As the amplitude of the applied AC increases, the reactance decreases
D. As the frequency of the applied AC increases, the reactance increases

Coils are effective in reducing alternator whine in high-frequency mobile installations. The higher the alternator whine frequency, the greater the reactance from the coil. Remember, *as f increases, X_L increases* ($X_L = 2\pi fL$). **ANSWER D**

G5A01 What is impedance?
A. The electric charge stored by a capacitor
B. The inverse of resistance
C. The opposition to the flow of current in an AC circuit
D. The force of repulsion between two similar electric fields

The term *impedance* means the *opposition to the flow of alternating current in a circuit*. Impedance to AC can be made up of resistance only, reactance only, or both resistance and reactance. You can create impedance to AC by winding a wire around a pencil to create a coil. This handy "choke" might minimize the alternator whine that may come in on your new worldwide mobile high-frequency station temporarily mounted in your vehicle. **ANSWER C**

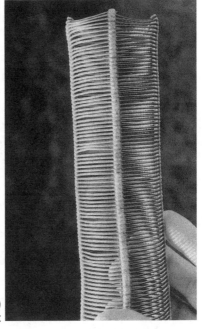

Coils Create Impedance (Opposition)
to the Flow of AC in a Circuit

G5A10 What unit is used to measure impedance?
A. Volt   C. Ampere
B. Ohm   D. Watt

The *ohm* is also used for measuring impedance. Thus, the ohm may mean impedance, reactance, or resistance. **ANSWER B**

G6A05 Which of the following is one effect of lead inductance in a capacitor used at VHF and above?
A. Effective capacitance may be reduced
B. Voltage rating may be reduced
C. ESR may be reduced
D. The polarity of the capacitor might become reversed

As a Technician Class operator preparing for your General Class ticket, you probably had a chance to look at VHF/UHF equipment circuit boards. Pieces of art, right? These rigs use SMT (surface mount technology) that places components directly on the board to minimize lead inductance. Conventional *capacitors with long*, scraggly *leads* would have *reduced effective capacitance*. **ANSWER A**

G6A06 What is the main disadvantage of using a conventional wire-wound resistor in a resonant circuit?
A. The resistor's tolerance value would not be adequate for such a circuit
B. The resistor's inductance could detune the circuit
C. The resistor could overheat
D. The resistor's internal capacitance would detune the circuit

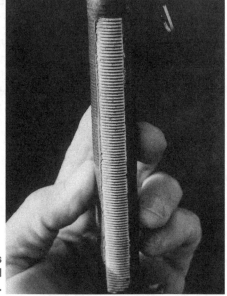

If you were able to scrape off the brown glaze coating on a *wire-wound resistor*, you would quickly see it looks exactly like a coil with evenly spaced turns. Actually, it IS an inductor and would not be suitable in any *circuit that could be detuned* accidentally through the use of the wrong component. Normally, we find wire-wound resistors in simple DC applications where we need to drop a small amount of voltage with relatively high current being passed. **ANSWER B**

A wire-wound resistor like this one is an inductor, and should not be used in tuned circuits.

G6A03 Which of the following is the primary advantage of ceramic capacitors?
A. Tight tolerance
B. High stability
C. High capacitance for given volume
D. Comparatively low cost

The ceramic capacitor is the workhorse in ham radio equipment. Their reliability is very good, and their *comparatively low cost* keeps ham radio equipment reasonably priced. **ANSWER D**

G6A07 What is an advantage of using a ferrite core with a toroidal inductor?
A. Large values of inductance may be obtained
B. The magnetic properties of the core may be optimized for a specific range of frequencies
C. Most of the magnetic field is contained in the core
D. All of these choices are correct

When you pass this test and earn General Class privileges, you may begin to operate High Frequency PORTABLE. A couple of rigs actually have batteries on the inside! With any portable transceiver, including VHF/UHF handhelds, MAKE

SURE THEY NEVER GET DROPPED. Dropping radio equipment could fracture the brittle iron cores within a toroidal inductor. The ferrite core fractures easily. The ferrite core within a toroidal inductor offers large values of inductance, with most of the magnetic field contained within the core so not to affect other components nearby. The toroidal inductor may be used in applications where core saturation IS desirable, so *all of the answer choices* to this question *are correct.* **ANSWER D**

G6A08 How should two solenoid inductors be placed so as to minimize their mutual inductance?
    A. In line with their winding axis
    B. With their winding axes parallel to each other
    C. With their winding axes at right angles to each another
    D. Within the same shielded enclosure
Solenoid inductors will interact if they are placed side by side, due to mutual inductance. To minimize this, 2 solenoid inductors should be *placed at right angles* to their winding axis to minimize unwanted mutual inductance. Just think of how a transformer winding FAVORS mutual inductance, but in this case, "at right angles" will MINIMIZE the effect. **ANSWER C**

G6A09 Why might it be important to minimize the mutual inductance between two inductors?
    A. To increase the energy transfer between both circuits
    B. To reduce or eliminate unwanted coupling
    C. To reduce conducted emissions
    D. To increase the self-resonant frequency of both inductors
If you ever had your radio opened up for a close look at all of the gizmos inside, you probably spotted several copper wire coils near the antenna output jack. Notice that none of these coils are parallel to each other. And same thing for any coils deep within the receiver or transmitter RF sections of your radio – most *coils are at right angles* to each other *to eliminate stray coupling* between the RF stages, or are placed in small cans to minimize mutual inductance. **ANSWER B**

G6A10 What is an effect of inter-turn capacitance in an inductor?
    A. The magnetic field may become inverted
    B. The inductor may become self resonant at some frequencies
    C. The permeability will increase
    D. The voltage rating may be exceeded
An inductor may be a small coil of wire, and in addition to offering a specific value of inductance, it may inadvertently offer inter-turn capacitance at some frequencies. This could cause the *coil to become self-resonant*, because we have now created both inductance and capacitance within a single component. **ANSWER B**

G5C01 What causes a voltage to appear across the secondary winding of a transformer when an AC voltage source is connected across its primary winding?
    A. Capacitive coupling   C. Mutual inductance
    B. Displacement current coupling   D. Mutual capacitance
Think of a transformer with interlaced coils. Through mutual inductance within the transformer, voltage applied to the primary will also appear across the secondary. *Mutual inductance.* **ANSWER C**

G5C02 Where is the source of energy normally connected in a transformer?

A. To the secondary winding   C. To the core

B. To the primary winding   D. To the plates

We normally hook the *primary winding* of a transformer to the source of energy. **ANSWER B**

G5C03 What is current in the primary winding of a transformer called if no load is attached to the secondary?

A. Magnetizing current   C. Excitation current

B. Direct current   D. Stabilizing current

When voltage is applied to the primary winding of the transformer and there is no load attached to the secondary, *magnetizing currents* will develop. **ANSWER A**

G5C06 What is the voltage across a 500-turn secondary winding in a transformer if the 2250-turn primary is connected to 120 VAC?

A. 2370 volts   C. 26.7 volts

B. 540 volts   D. 5.9 volts

This is a *turns ratio problem*, and is relatively easy to solve using the following equation:

$$E_S = E_P \times \frac{N_S}{N_P} = \frac{E_P \times N_S}{N_P}$$

which means the *voltage of the secondary is equal to the voltage of the primary times the number of turns of the secondary divided by the number of turns of the primary*. It is derived from the equation that says that the ratio of the secondary voltage, E_S, to the primary voltage, E_P, is equal to the ratio of the turns on the secondary, N_S, to the turns on the primary, N_P.

$$\frac{E_S}{E_P} = \frac{N_S}{N_P}$$

Multiply 120 (E_P) times 500 (N_S), and then divide your answer by 2250. This gives you *26.7 volts*. Calculator keystrokes are: Clear, 120 x 500 ÷ 2250 = 26.7 volts. **ANSWER C**

G5C07 What is the turns ratio of a transformer used to match an audio amplifier having a 600-ohm output impedance to a speaker having a 4-ohm impedance?

A. 12.2 to 1   C. 150 to 1

B. 24.4 to 1   D. 300 to 1

The equation that applies is:

$$\frac{N_P}{N_S} = \sqrt{\frac{Z_P}{Z_S}} \quad \text{derived from} \quad \frac{Z_P}{Z_S} = \left(\frac{I_S}{I_P}\right)^2 = \left(\frac{N_P}{N_S}\right)$$

The ratio of the turns on the primary, N_P, to the turns on the secondary, N_S, is equal to the square root of the ratio of the primary impedance, Z_P, to the secondary impedance, Z_S. Remember that this turns ratio is primary to secondary. The ratio in question G5C06 is secondary to primary.

Don't worry if you have forgotten about square roots. There's an easy way to solve the problem. The primary impedance, Z_P, of the transformer must match the

600-ohms output impedance of the amplifier; therefore, Z_P is 600 ohms. Divide 600 ohms by 4 ohms, the speaker load impedance on the secondary, and you end up with 150.

Now you need to find the square root of 150. You know that a square root multiplied by itself gives you the number you want. You can do it by approximation. Since 12 x 12 = 144 and 13 x 13 = 169, you know that the square root of 150 is between 12 and 13. The only answer given that is close is 12.2. Choose it and you have the correct answer. See, you didn't have to remember how to do square roots.

The *calculator keystrokes are:* Clear, *600 ÷ 4 x 150*, then press the *square root key* to produce the answer, *12.25.* **ANSWER A**

G5A11 Why should core saturation of a conventional impedance matching transformer be avoided?
A. Harmonics and distortion could result
B. Magnetic flux would increase with frequency
C. RF susceptance would increase
D. Temporary changes of the core permeability could result

If you are into data communications, there is plenty of software and hardware to marry your ham radio and computer together. A critical connection will be data out and data in between the radio and computer, and impedance matching transformers are many times found in that magic controller box from MFJ or the West Mountain Radio Rigrunner (TM). A peek inside will reveal a couple of small impedance-matching transformers. They are designed so that increasing the current through the coil will cause an increase of proportional flux. However, too much current within the transformer primary may saturate the core with no appreciable change or increase in the flux value. The saturation point is dependent on the materials found inside that tiny little transformer. When the flux is forced beyond its rated value, permeability of the core decreases, and DISTORTION may occur if you're transmitting or distortion may occur if you are receiving with your volume control turned up too high. *Distortion and harmonics* are to be avoided! **ANSWER A**

Website Resources

▼ IF YOU'RE LOOKING FOR	▼ THEN VISIT
Learn More about Electricity · · · · · · · ·	www.forrestmims.com
	www.masterpublishing.com
	www.w5yi.org
	www.arrl.org

G4C06 Which of the following is an important reason to have a good station ground?

A. To reduce the likelihood of RF burns
B. To reduce the likelihood of electrical shock
C. To reduce interference
D. All of these answers are correct

It is important to have good grounding of your new high frequency General Class station equipment. A good earth ground on your home HF station will give you a "cold mic" where the metal front of the microphone won't burn your ruby red lips. A good ground will also reduce the chance of electrical shock and will help reduce interference to your home electronics. *All of these are important reasons to have a good station ground.* **ANSWER D**

G4C05 What might be the problem if you receive an RF burn when touching your equipment while transmitting on an HF band, assuming the equipment is connected to a ground rod?

A. Flat braid rather than round wire has been used for the ground wire
B. Insulated wire has been used for the ground wire
C. The ground rod is resonant
D. The ground wire is resonant

A *33-foot ground wire becomes self-resonant* on both 40 meters and 15 meters, giving you the chance of getting an RF burn as you hold the metal mic or touch the metal panel on your radio. Always use ground foil on long ground runs to minimize this problem. **ANSWER D**

G4C09 Which of the following statements about station grounding is true?

A. The chassis of each piece of station equipment should be tied together with high-impedance conductors

B. If the chassis of all station equipment is connected with a good conductor, there is no need to tie them to an earth ground

C. RF hot spots can occur in a station located above the ground floor if the equipment is grounded by a long ground wire

D. A ground loop is an effective way to ground station equipment

A good second story station ground is wide copper foil going down to a good moist earth ground connection and ground rod. Avoid using wires, as *long ground wires* will sometimes get *"RF hot"* and reduce the capacity of your ground system. **ANSWER C**

Copper Foil Ground Strap Provides
a Good Surface Area Ground

GROUND FOIL FACTS:
Receive a FREE grounding info package plus a 3˝ sample of copper foil by sending an e-mail request to David at Metal-Cable, Inc. His e-mail address is: **david@metal-cable.com**.

G4C07 What is one good way to avoid stray RF energy in an amateur station?

A. Keep the station's ground wire as short as possible

B. Install an RF filter in series with the ground wire

C. Use a ground loop for best conductivity

D. Install a few ferrite beads on the ground wire where it connects to your station

If you run *ground wires*, as opposed to copper foil, keep those wires *as short as possible*. **ANSWER A**

G4C13 How can a ground loop be avoided?

A. Series connect all ground conductors

B. Connect the AC neutral conductor to the ground wire

C. Avoid using lock washers and star washers in making ground connections

D. Connect all ground conductors to a single point

I have several HF transceivers at the shack, and all ground copper foil connections go to a single ground point. There the flattened copper pipe goes through the wall and into the earth. *Grounding to a single point will minimize ground loops.* If ever you had a new fancy audio stereo system installed in your vehicle, the savvy tech runs all component grounds to a single ground point to minimize ground loops. **ANSWER D**

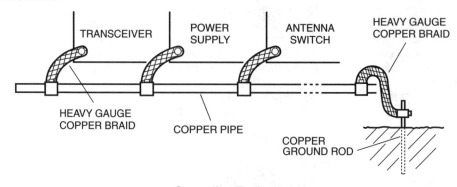

Grounding Equipment

G4C01 Which of the following might be useful in reducing RF interference to audio-frequency devices?
A. Bypass inductor
B. Bypass capacitor
C. Forward-biased diode
D. Reverse-biased diode

When you begin operating on General Class frequencies, your powerful, high-frequency SSB transceiver fed into a roof-top antenna system will probably create audio-frequency interference to your own home electronics, and those of your surrounding four neighbors. Bypass capacitors – usually 0.01 mF – will sometimes help minimize this problem when strategically placed across and onto speaker wires and wiring harnesses inside the affected home electronic systems. It's not a cure-all, but *bypass capacitors* are your first step in *resolving interference* complaint problems on a case-by-case basis. **ANSWER B**

G4C08 Which of the following is a reason to place ferrite beads around audio cables to reduce common mode RF interference?
A. They act as a series inductor
B. They act as a shunt capacitor
C. They lower the impedance of the cable
D. They increase the admittance of the cable

You can purchase ferrite chokes and beads for your hi-fi wires to help reduce signals from getting into your audio system downstairs. The *ferrite beads* and chokes *act like a series indictor* that traps RF energy. **ANSWER A**

G4C02 Which of the following should be installed if a properly operating amateur station is interfering with a nearby telephone?
A. An RFI filter on the transmitter
B. An RFI filter at the affected telephone
C. A high pass filter on the transmitter
D. A high pass filter at the affected telephone

Telephone RFI (radio frequency interference) filters may need to be installed at the drop point where the phone lines enter the house. The phone company may install these filters at no charge; however, you will probably need to buy additional *filters* to be *placed* in series *with the phone line jacks* inside the house. Phone companies might also do this, but for a charge. **ANSWER B**

G4C03 What sound is heard from a public-address system if there is interference from a nearby single-sideband phone transmitter?
A. A steady hum whenever the transmitter is on the air
B. On-and-off humming or clicking
C. Distorted speech
D. Clearly audible speech

Single-sideband sounds like *distorted speech* coming over a public address system or certain home electronics. However, AM CB radio transmissions usually come through loud and clear, so these are easily distinguished from SSB ham transmissions. If someone says you are causing interference, ask the big question: "Does it sound garbled, or does it sound clear?" **ANSWER C**

G4C04 What is the effect on a public-address system if there is interference from nearby CW transmitter?
A. On-and-off humming or clicking
B. A CW signal at a nearly pure audio frequency
C. A chirpy CW signal
D. Severely distorted audio

CW transmissions come over a PA or home electronics system as *on-and-off humming or clicking* sounds. There is no mistaking the sound of CW. **ANSWER A**

G4C11 Which of the following can cause unintended rectification of RF signal energy and can result in interference to your station as well as nearby radio and TV receivers?
A. Induced currents in conductors that are in poor electrical contact
B. Induced voltages in conductors that are in good electrical contact
C. Capacitive coupling of the RF signal to ground
D. Excessive standing wave ratio (SWR) of the transmission line system

The other day I was yakking on 20 meters when my hard-of-hearing neighbor next door called on the phone and said I was getting into her hearing aid. My appropriate ham radio response was to blame it on the CB radio operator next door, but I offered to look over her hearing apparatus to see if I could figure out what was going on. Her specific "hearing enhancer" (not advertised as a hearing aid) purchased from a catalog company features a miniature plug that goes into a little plastic box hanging around her neck. She told me the plug connection was noisy, too, as well as most recently picking up the radio station downtown! A little bit of contact cleaner and the problem disappeared. *Induced CURRENTS* in the hearing enhancer wires were picking up the signal, *and the poor electrical contact* at the plug was creating unintended rectification. Now, as for the guy across the street picking up my signals in his dental fillings... I don't know. **ANSWER A**

G4C12 What is one cause of broadband radio frequency interference at an amateur radio station?
   A. Not using a balun or line isolator to feed balanced antennas
   B. Lack of rectification of the transmitter's signal in power conductors
   C. Arcing at a poor electrical connection
   D. The use of horizontal, rather than vertical antennas

About once every 6 months I go around and tighten up all connections on my ham equipment. Most important is the tightening up of the copper ground foil connections to the back of my rig. *An intermittent RF ground* will sometimes create *broadband radio frequency noise* that will magically go away as soon as you give that nut a little clockwise crank. **ANSWER C**

G4E02 What is alternator whine?
   A. A DC emission from the alternator
   B. A constant pitched tone or buzz in transmitted or received audio that occurs whenever the ignition key is in the on position
   C. A tone or buzz in transmitted or received audio that varies with engine speed
   D. A mechanical sound from the alternator indicating current overload

Alternator whine on your new HF mobile is sometimes detected by the other station when you are transmitting. They will say your signal has a *tone* on it *that changes in pitch as you speed up and slow down*. This indeed is alternator whine, and in-line filters from your car stereo store designed for 20 amps of current will generally solve the problem completely. **ANSWER C**

G3C14 Which of the following antennas will be most effective for skip communications on 40 meters during the day?
A. A vertical antenna
B. A horizontal dipole placed between 1/8 and 1/4 wavelength above the ground
C. A left-hand circularly polarized antenna
D. A right-hand circularly polarized antenna

A simple *dipole antenna*, placed between 1/8 and 1/4 wavelength above the ground, will give you powerful daytime skywave communications *on the 40 meter band*. To reach out further, elevate it. To "pull in" your first hop, slightly lower it!
ANSWER B

G9B10 What is the approximate length for a 1/2-wave dipole antenna cut for 14.250 MHz?

A. 8.2 feet   C. 24.6 feet
B. 16.4 feet   D. 32.8 feet

To calculate the length, in feet, of a half-wavelength dipole antenna, divide 468 by the antenna's operating frequency in MHz. Just remember: 2-4-6-8, Who do we appreciate? Calculators ARE PERMITTED for your examinations, so here are the keystrokes to divide 468 by the stated test question frequency: Clear, clear, 4-6-8 ÷ 14.250 = And your answer comes out 32.842 feet, answer D. Yikes! Constructing this dipole, each side will be about *16.4 feet* long, and how do you go from a

---

decimal point feet to inches? No problem – multiply the decimal feet by 12 and you end up with inches. Remember, the half wave dipole needs the center insulator to go exactly in the middle. Each side would then be 1/4 wavelength. Again, the formula for a 1/2 wavelength dipole is:

$$\frac{468}{fMHz} = \text{wavelength in feet. } \textbf{ANSWER D}$$

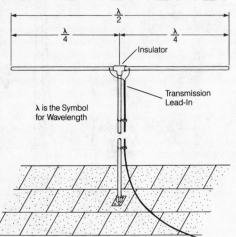

Elmer Point:
The half-wavelength dipole antenna offers no compromise when transmitting and receiving long-range signals. A half-wave dipole usually will outperform multi-band medium length verticals and excel above a fancy $800 mobile, big-coil whip. Dipoles are simple to construct and you can pull up the center section with a rope to configure it as an inverted vee. The only thing that beats a dipole is a big 3-element beam or a quad.

G9B11 What is the approximate length for a 1/2-wave dipole antenna cut for 3.550 MHz?
A. 42.2 feet
B. 84.5 feet
C. 131.8 feet
D. 263.6 feet
Let's do the math: 468 ÷ 3.55 = *131.8 feet!* Easy, huh? **ANSWER C**

G9B09 Which of the following is an advantage of a horizontally polarized as compared to vertically polarized HF antenna?
A. Lower ground reflection losses
B. Lower feed-point impedance
C. Shorter Radials
D. Lower radiation resistance
The well constructed and elevated 1/2 wave dipole will generally outperform ground mounted verticals. This is because the ground mounted vertical usually encounters *ground reflection losses* with everything close by to its base. **ANSWER A**

G9B07 How does the feed-point impedance of a 1/2 wave dipole antenna change as the antenna is lowered from 1/4 wave above ground?
A. It steadily increases
B. It steadily decreases
C. It peaks at about 1/8 wavelength above ground
D. It is unaffected by the height above ground
The 1/2 wave dipole is an outstanding antenna system that will cost you quarters to build. The 1/2 wave dipole, elevated to roof level, will generally outperform expensive high frequency mobile whips, expensive trap vertical antennas, and outdo expensive automatic tuner fed long wires. But get the 1/2 wave dipole up high. If it is less than 1/4 wavelength above ground, the *feed point impedance will dramatically decrease*, and your radiation pattern goes whacko. **ANSWER B**

G9B08 How does the feed-point impedance of a 1/2 wave dipole change as the feed-point location is moved from the center toward the ends?
A. It steadily increases
B. It steadily decreases
C. It peaks at about 1/8 wavelength from the end
D. It is unaffected by the location of the feed-point

If you try to offset the feed point of a 1/2 wave dipole from the center to the ends, *feed point impedance will rise* well above 50 ohms, and your rig's automatic antenna tuner may have a hard time bringing the SWR down to 1:1. **ANSWER A**

G9B04 What is the low angle azimuthal radiation pattern of an ideal half-wavelength dipole antenna installed 1/2 wavelength high and parallel to the earth?
A. It is a figure-eight at right angles to the antenna
B. It is a figure-eight off both ends of the antenna
C. It is a circle (equal radiation in all directions)
D. It has a pair of lobes on one side of the antenna and a single lobe on the other side

The dipole gives you a *figure eight pattern* at right angles to the antenna wire. Reception and transmission are minimal off the ends of the wire. **ANSWER A**

G9B05 How does antenna height affect the horizontal (azimuthal) radiation pattern of a horizontal dipole HF antenna?
A. If the antenna is too high, the pattern becomes unpredictable
B. Antenna height has no effect on the pattern
C. If the antenna is less than 1/2 wavelength high, the azimuthal pattern is almost omnidirectional
D. If the antenna is less than 1/2 wavelength high, radiation off the ends of the wire is eliminated

Never mount a dipole antenna less than one-half wavelength from the earth for long-range DX. If you do, you may have signal distortion, an *omni-directional radiation pattern*, and most of your signal going straight up. **ANSWER C**

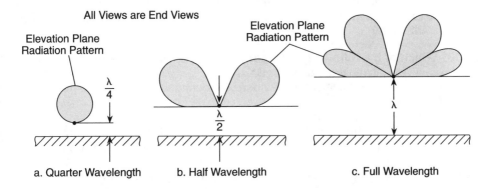

The Radiation Pattern of an Antenna Changes as Height Above Ground is Varied
Source: *Antennas*, A.J. Evans, K.E. Britain, © 1998, Master Publishing, Niles, Illinois

---

G3C13 What is Near Vertical Incidence Sky-wave (NVIS) propagation?
A. Propagation near the MUF
B. Short distance HF propagation using high elevation angles
C. Long path HF propagation at sunrise and sunset
D. Double hop propagation near the LUF

Let's imagine you going out on a camping trip into a valley 200 miles away from your buddy's house. With a normal dipole up about a half wave length, your 40 meter signal is literally skipping over his QTH. A neat way to "pull in" your first skywave hop is to lower your dipole close to the ground just high enough so no one can touch it. You will find that your first skywave hop now arrives at your friend's house loud and clear. Dramatically *lowering an antenna* on purpose close to the Earth *will develop a higher elevation angle* of radiation, *causing* your skip distance to "pull in" to establish *shorter-than-normal skywave communications*.
ANSWER B

G9D01 What does the term "NVIS" mean as related to antennas?
A. Nearly Vertical Inductance System
B. Non-Visible Installation Specification
C. Non-Varying Impedance Smoothing
D. Near Vertical Incidence Skywave

The dipole antenna is fun to play with when strung between two trees with two pulleys. String it way up high, and your angle of radiation lowers, so you get a longer distance skip. However, if you want to talk with your Granddad who is about 300 miles away, try lowering the dipole real close to the ground and notice that the distant skywave skip will fade out and good old Granddad's signal just a couple hundred miles away now comes in stronger. This is because your *skywave is nearly vertical.* **ANSWER D**

G9D03 At what height above ground is an NVIS antenna typically installed?
A. As close to one-half wave as possible
B. As close to one wavelength as possible
C. Height is not critical as long as significantly more than 1/2 wavelength
D. Between 1/10 and 1/4 wavelength

To get your dipole to operate like a *NVIS antenna*, start near ground level and work from *1/10 to 1/4 wavelength* and listen to close in stations get dramatically stronger. **ANSWER D**

G9D02 Which of the following is an advantage of an NVIS antenna?
A. Low vertical angle radiation for DX work
B. High vertical angle radiation for short skip during the day
C. High forward gain
D. All of these choices are correct

The military and emergency communicators will use NVIS dipoles just a few feet above the ground to cause their signals to take an almost *vertical takeoff* to the ionosphere, and come back down much closer in than a regular dipole up one half wavelength. This is great for *short skip during the day*. **ANSWER B**

G9B12 What is the approximate length for a 1/4-wave vertical antenna cut for 28.5 MHz?
A. 8.2 feet
B. 10.5 feet
C. 16.4 feet
D. 21.0 feet

Let's say you want to operate 10 meters mobile. When you operate mobile, the metal frame of the vehicle makes up 1/4 wavelength of your antenna system, and the other *1/4 wavelength* is the VERTICAL radiating antenna. Think of your vertical whip as one side of a dipole and the vehicle chassis as the other. First calculate a half wavelength for *28.5 MHz*... 4-6-8 ÷ 28.5 =16.42. Now divide 16.42 by 2 to end up with a quarter whip, *8.2 feet*. Sure, you could use "2-3-4" instead of "4-6-8", but I like "4-6-8" because even a mobile antenna is a form of a 1/2 wave dipole. The formula for a 1/4 wave is:

$$\frac{234}{fMHz} = 1/4 \text{ wave in feet. } \textbf{ANSWER A}$$

G9B03 What happens to the feed-point impedance of a ground-plane antenna when its radials are changed from horizontal to downward-sloping?
   A. It decreases
   B. It increases
   C. It stays the same
   D. It reaches a maximum at an angle of 45 degrees
The feedpoint *impedance increases* from 30 ohms to 50 ohms when you bend the radials downward. **ANSWER B**

G9B02 What is an advantage of downward sloping radials on a ground-plane antenna?
   A. They lower the radiation angle
   B. They bring the feed-point impedance closer to 300 ohms
   C. They increase the radiation angle
   D. They can be adjusted to bring the feed-point impedance closer to 50 ohms

On a simple ground-plane antenna, the *radials are bent down* about 45 degrees to *bring the feedpoint impedance close to 50 ohms*. If they stick straight out, the impedance is more like 30 ohms, causing a mismatch.
ANSWER D

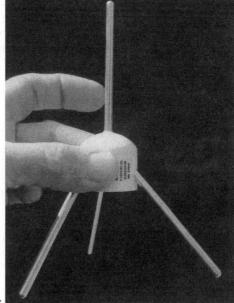

A Ground-Plane Antenna.

G9B06 Where should the radial wires of a ground-mounted vertical antenna system be placed?
A. As high as possible above the ground
B. Parallel to the antenna element
C. On the surface or buried a few inches below the ground
D. At the top of the antenna

Ground radial wires on the surface or buried just below the surface are important for the ground-mounted vertical antenna to establish its own counterpoise. To hear any real difference when you already have a few radials laid out per band, you must double the number of radials. If you have 2 per band, try 4. If you have 4, try 8. If you need more ground plane than 8 radials per band, try mounting your antenna on the aluminum shed in the backyard. The more ground radials you have on a ground plane antenna, the lower the takeoff angle of radiation. And this means more DX. Get out the shovel, and start digging to install more ground radials! **ANSWER C**

G2D11 Which HF antenna would be the best to use for minimizing interference?
A. A bi-directional antenna
B. An isotropic antenna
C. A unidirectional antenna
D. An omnidirectional antenna

Here is a correct answer that is unnecessarily disguised. The term "uni" means single, and the correct answer of a "unidirectional antenna" focusing the signal to minimize interference in an alternate direction is indeed the correct answer. We normally call a *"unidirectional antenna"* a Yagi, or a beam, or simply directional – but UNIdirectional? **ANSWER C**

G9B01 What is one disadvantage of a directly fed random-wire antenna?
A. It must be longer than 1 wavelength
B. You may experience RF burns when touching metal objects in your station
C. It produces only vertically polarized radiation
D. It is not effective on the higher HF bands

Unless you use a remote-mounted automatic antenna tuner, the *random-wire antenna* can put a lot of *RF feedback* in your station. **ANSWER B**

G9C18 Which of the following antenna types consists of a driven element and some combination of parasitically excited reflector and/or director elements?
A. A collinear array   C. A double-extended Zepp antenna
B. A rhombic antenna   D. A Yagi antenna

The term excited does not refer to your Grandpa when you tell him you just passed your General Class ticket. Rather, *parasitically excited* refers to those *elements* hanging on the boom of a *Yagi*, such as directors and reflectors with no apparent coaxial cable connection. The coax only goes to the driven element, and all of the other reflectors and directors are doing their work just hanging on the boom, insulated or not, with no direct coax connection! **ANSWER D**

G9C04 Which statement about a Yagi antenna is true?

A. The reflector is normally the longest parasitic element
B. The director is normally the longest parasitic element
C. The reflector is normally the shortest parasitic element
D. All of the elements must be the same length

You can always figure out which way to point a 3 element Yagi by seeing which elements on the boom are the shortest. *The reflector is normally the longest parasitic element*, and one or more directors are usually 5 % shorter than the driven element. So scope it out, and the shortest elements of the Yagi always point in the general direction of the distant station. Most worldwide Yagis are polarized horizontal, too. **ANSWER A**

G9C02 What is the approximate length of the driven element of a Yagi antenna?

A. 1/4 wavelength
B. 1/2 wavelength
C. 3/4 wavelength
D. 1 wavelength

Since the *Yagi antenna is a series of dipoles* affixed to a boom in the same plane, the individual *elements are about one-half wavelength long,* just like the driven element. The reflector is a little longer, the director is a little shorter. **ANSWER B**

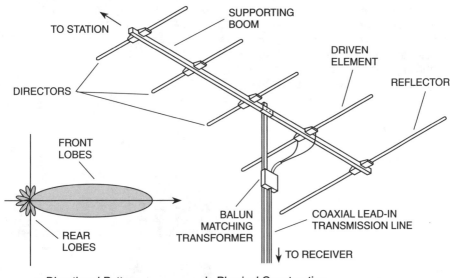

a. Directional Pattern   b. Physical Construction

A Beam Antenna - The Yagi Antenna

Source: *Antennas - Selection and Installation,*© 1986 Master Publishing, Inc., Niles, IL

G9C03 Which statement about a three-element single-band Yagi antenna is true?

A. The reflector is normally the shortest parasitic element
B. The director is normally the shortest parasitic element
C. The driven element is the longest parasitic element
D. Low feed-point impedance increases bandwidth

On a three-element beam, *the director is shorter* than the driven element, and the reflector is longer than the driven element. **ANSWER B**

G9C06 Which of the following is a reason why a Yagi antenna is often used for radio communications on the 20 meter band?
- A. It provides excellent omnidirectional coverage in the horizontal plane
- B. It is smaller, less expensive and easier to erect than a dipole or vertical antenna
- C. It helps reduce interference from other stations to the side or behind the antenna
- D. It provides the highest possible angle of radiation for the HF bands

When you pass your General Class license, the *Yagi antenna* will give you one of the best signals ever on the *20-meter band* because it *reduces interference* from other stations off to the side or behind. A small, three-element Yagi works quite nicely just a few feet off the top of your house. **ANSWER C**

G9C05 What is one effect of increasing the boom length and adding directors to a Yagi antenna?
- A. Gain increases
- B. SWR increases
- C. Weight decreases
- D. Wind load decreases

On a worldwide (as well as VHF/UHF) *Yagi antenna, boom length determines the amount of gain*. The number of elements and the diameter of the elements influence the directivity and bandwidth of the antenna, but the big factor that determines which Yagi antenna is going to outperform another Yagi in gain is boom length. **ANSWER A**

A "Mobile" CushCraft 5-Band, 3-Element Yagi in Operation at a Field Day Event.

G9C08 What is meant by the "main lobe" of a directive antenna?
- A. The magnitude of the maximum vertical angle of radiation
- B. The point of maximum current in a radiating antenna element
- C. The maximum voltage standing wave point on a radiating element
- D. The direction of maximum radiated field strength from the antenna

The *main lobe* is the *main radiating direction* of the signal. You may use a field strength meter to determine the main lobe of most Yagi antennas. **ANSWER D**

G9C07 What does "front-to-back ratio" mean in reference to a Yagi antenna?
A. The number of directors versus the number of reflectors
B. The relative position of the driven element with respect to the reflectors and directors
C. The power radiated in the major radiation lobe compared to the power radiated in exactly the opposite direction
D. The ratio of forward gain to dipole gain

The Yagi antenna gives you an excellent front-to-back ratio. This means that the *majority of the power is radiated out of the front* of the antenna, with *little signal wasted to the back* or to the sides. **ANSWER C**

G9C09 What is the approximate maximum theoretical forward gain of a 3 Element Yagi antenna?
A. 9.7 dBi
B. 7.3 dBd
C. 5.4 times the gain of a dipole
D. All of these choices are correct

This is a great question with a much-debated answer – What's the most gain we can get out of a 3 element Yagi? There are plenty of variables including boom length, element construction, feed point techniques… but to get this answer correct, go for answer A, 9.7 dBi – 9.7 decibels refers to an isotropic antenna which has 0 dB gain. **ANSWER A**

G9C10 Which of the following is a Yagi antenna design variable that could be adjusted to optimize forward gain, front-to-back ratio, or SWR bandwidth?
A. The physical length of the boom
B. The number of elements on the boom
C. The spacing of each element along the boom
D. All of these choices are correct

Any time you begin moving elements along the boom of a Yagi, gain, front to back ratios, and SWR bandwidth will be affected. *So, all of the choices are correct* about the boom length and the number and spacing of the elements along the boom. **ANSWER D**

G9C01 How can the SWR bandwidth of a Yagi antenna be increased?
A. Use larger diameter elements
B. Use closer element spacing
C. Use traps on the elements
D. Use tapered-diameter elements

The *greater the diameter of the elements, the greater the bandwidth* of a Yagi beam antenna for worldwide operation. This is why a wire beam antenna does not offer as much bandwidth as one constructed of large aluminum tubes. **ANSWER A**

G9C11 What is the purpose of a "gamma match" used with Yagi antennas?
A. To match the relatively low feed-point impedance to 50 ohms
B. To match the relatively high feed-point impedance to 50 ohms
C. To increase the front to back ratio
D. To increase the main lobe gain

If you decide to construct your own 3 element beam, or if your Granddad gives you his old 3 element 20 meter monobander, chances are you will need to develop a feed point connection that will *take the relatively LOW feed point impedance up to 50 ohms to match that of your coax.* The gamma match is a series line matching system that lets you tap the driven element at a preferred 50 ohm point. To an ohm meter, both the gamma match and the T match may look like a short circuit – but to your radio frequency signal on the 20 meter band for this monoband beam, it will look like a perfect match! **ANSWER A**

G9C12 Which of the following describes a common method for insulating the driven element of a Yagi antenna from the metal boom when using a gamma match?
A. Support the driven element with ceramic standoff insulators
B. Insert a high impedance transformer at the driven element
C. Insert a high voltage balun at the driven element
D. None of these answers are correct. No insulation is needed

The gamma match is a favorite among hams who build their own beams because you do not need to go through the work of building element insulators to try to isolate the driven element from the boom. *With a gamma match, no insulation is needed!* **ANSWER D**

G9D05 What is the advantage of vertical stacking of horizontally polarized Yagi antennas?
A. Allows quick selection of vertical or horizontal polarization
B. Allows simultaneous vertical and horizontal polarization
C. Narrows the main lobe in azimuth
D. Narrows the main lobe in elevation

Vertically stacking horizontal Yagis *helps concentrate the main lobe in ELEVATION* for added signal strength to the desired station. Aiming (beamwidth) the antenna with your rotator remains about the same. **ANSWER D**

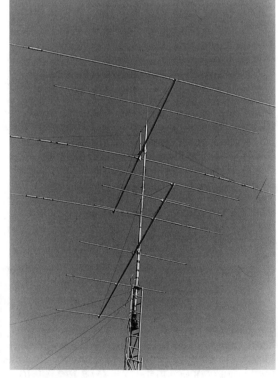

High frequency beam antennas, on different bands, may be stacked vertically, with little interaction, for a clean look and solid performance. Try for maximum seperation to keep the radiation patterns clean.

G9D04 How does the gain of two 3-element horizontally polarized Yagi antennas spaced vertically 1/2 wave apart from each other typically compare to the gain of a single 3-element Yagi?
    A. Approximately 1.5 dB higher   C. Approximately 6 dB higher
    B. Approximately 3 dB higher   D. Approximately 9 dB higher
You don't need a linear amplifier to boost your effective radiated power output, plus increase reception. *Stack a pair of Yagis* about one half wavelength apart and pick up a 2 times *increase* in transmit and receive *gain. (2x = 3 dB)* **ANSWER B**

G9D12 What is the primary purpose of traps installed in antennas?
    A. To permit multiband operation
    B. To notch spurious frequencies
    C. To provide balanced feed-point impedance
    D. To prevent out of band operation
A great way to get started on 40 meters, 20 meters, 15 meters, and 10 meters is with a trap dipole or a trap beam antenna. You could also do a trap vertical antenna on the roof. The *traps* will allow the *multiband antenna* to naturally tune to the specific ham bands. Each trap acts like a stop-band for longer wavelengths, yet offers series loading when you specifically want to operate on longer wavelength lower frequencies. I operate a 4-element, 4-band beam antenna and the traps work great to give me high frequency operation on 4 specific ham bands. When you get your new General license, let's try to work each other on the air. **ANSWER A**

G9D11 Which of the following is a disadvantage of multiband antennas?
    A. They present low impedance on all design frequencies
    B. They must be used with an antenna tuner
    C. They must be fed with open wire line
    D. They have poor harmonic rejection
Newer HF ham radios have internal band pass filters to minimize harmonics. A harmonic is multiple of your fundamental frequency. If you are transmitting on 7 MHz, you would want to minimize any harmonic coming out on 14 MHz to nearby hams on that frequency. Unfortunately, the many advantages of a multiband antenna also incur the disadvantage of *poor harmonic rejection.* **ANSWER D**

G9C19 What type of directional antenna is typically constructed from 2 square loops of wire each having a circumference of approximately one wavelength at the operating frequency and separated by approximately 0.2 wavelength?
    A. A stacked dipole array   C. A cubical quad antenna
    B. A collinear array   D. An Adcock array
Square loops of wire, a total of one wavelength in wire length, must surely be the *cubicle quad antenna.* It is a monster. It hates rain and ice! **ANSWER C**

G9C14 How does the forward gain of a 2-element cubical-quad antenna compare to the forward gain of a 3 element Yagi antenna?
    A. 2/3   C. 3/2
    B. About the same   D. Twice
If you live in the northeast, severe icing might quickly destroy the 2 element cubicle quad antenna constructed of wire and fiberglass spreaders. If you get a lot of wind, snow, and ice, best consider a 3 element Yagi over the 2 element cubicle quad. *The gain is about the same.* **ANSWER B**

G9C13 Approximately how long is each side of a cubical-quad antenna driven element?

    A. 1/4 wavelength   C. 3/4 wavelength
    B. 1/2 wavelength   D. 1 wavelength

The cubical-quad antenna is a full-wavelength, four-sided, wire antenna system that offers identical to slightly improved performance over a Yagi. Each side of the cubicle quad is 1/4 wavelength. This makes *the cubicle quad a full wavelength antenna*. To determine the length of wire for each side of the driven element, use the following formula:

$$\text{Driven Element for each side (in feet)} = \frac{1005}{f\ (\text{MHz})} \div 4$$

ANSWER A

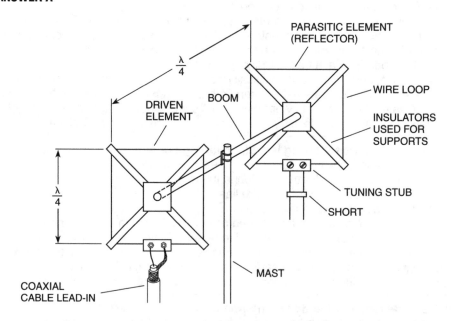

A Two-Element Cubical Quad Antenna - Horizontally Polarized
Source: *Antennas—Selection and Installation*, © 1986 Master Publishing, Inc., Niles, IL

G9C21 What configuration of the loops of a cubical-quad antenna must be used for the antenna to operate as a beam antenna, assuming one of the elements is used as a reflector?

    A. The driven element must be fed with a balun transformer
    B. The driven element must be open-circuited on the side opposite the feed-point
    C. The reflector element must be approximately 5% shorter than the driven element
    D. The reflector element must be approximately 5% longer than the driven element

Remember, *reflectors are generally 5% longer than the driven element* and directors are 5% shorter. They ask about the reflector, so it will be 5% longer than the driven element. **ANSWER D**

G9C15 Approximately how long is each side of a cubical-quad antenna reflector element?

    A. Slightly less than 1/4 wavelength
    B. Slightly more than 1/4 wavelength
    C. Slightly less than 1/2 wavelength
    D. Slightly more than 1/2 wavelength

Here they want to know the length of each side of the *reflector* element, which is always going to be a bit *longer* than the driven element. Use the same formula as for the previous question but substitute 1030 for 1005, then divide by the frequency in MHz. Remember, divide your answer by 4 because they only want to know the dimension of one side. **ANSWER B**

G9C20 What happens when the feed-point of a cubical quad antenna is changed from the center of the lowest horizontal wire to the center of one of the vertical wires?

    A. The polarization of the radiated signal changes from horizontal to vertical
    B. The polarization of the radiated signal changes from vertical to horizontal
    C. The direction of the main lobe is reversed
    D. The radiated signal changes to an omnidirectional pattern

What a work of art – you have built your own cubicle quad, strung between two coconut trees, with each side 1/4 wavelength long for the band of choice. If you feed coax cable to the center of the horizontal side, the polarization will be horizontal to the earth, and this is called... you guessed it... horizontal polarization. If you decide to *feed the quad* in the center of the *vertical side*, guess what? The radiation will leave the *antenna vertically polarized*. Will all this work of changing from the horizontal to the vertical make any big difference via skywaves? Probably not. **ANSWER A**

G9C16 How does the gain of a two element delta-loop beam compare to the gain of a two element cubical quad antenna?

    A. 3 dB higher   C. 2.54 dB higher
    B. 3 dB lower   D. About the same

The Delta loop beam may look like a triangle with each side 1/3 wavelength long. When compared to the 4 sided cubicle quad antenna, their *gain performance is almost identical.* **ANSWER D**

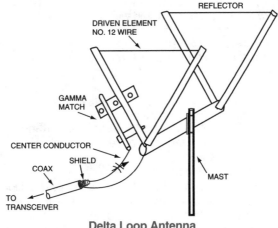

Delta Loop Antenna

G9C17 Approximately how long is each leg of a symmetrical delta-loop antenna driven element?
    A. 1/4 wavelengths   C. 1/2 wavelengths
    B. 1/3 wavelengths   D. 2/3 wavelengths

The Delta-loop is another version of a full-wavelength antenna system. The beauty of the delta-loop is you can put the apex of the loop way up at the top of a tree. *Each side of the symmetrical delta-loop antenna is 1/3 wavelengths long.* Use the formula:

$$\text{Driven Element for each side (in feet)} = \frac{1005}{f\,(MHz)} \div 3$$

Just as you did for the quad, divide 1005 by the frequency in MHz. However, since this is a delta-loop, and they only want to know one side, divide your answer by 3 instead of 4. **ANSWER B**

G9D07 Which of the following describes a log periodic antenna?
    A. Length and spacing of the elements increases logarithmically from one end
        of the boom to the other
    B. Impedance varies periodically as a function of frequency
    C. Gain varies logarithmically as a function of frequency
    D. SWR varies periodically as a function of boom length

They call this monster antenna a "logarithmic" periodic antenna because the *length and spacing of the elements increase logarithmically* from one end of the boom to the other. But they are monster antennas with limited applications for specific ham radio bands. Best go with a multiband dipole or a multiband beam, long before you ever consider the monster "log"! **ANSWER A**

G9D06 Which of the following is an advantage of a log periodic antenna?
    A. Wide bandwidth
    B. Higher gain per element than a Yagi antenna
    C. Harmonic suppression
    D. Polarization diversity

Some military stations may use an antenna system called a *"log periodic."* The elements get progressively shorter on a very long boom, allowing for a *wide bandwidth* over many MHz of radio bands. SWR remains flat on a high frequency log periodic antenna from 10 MHz all the way up to 50 MHz! The only problem with a log antenna is the low forward gain achieved with such a monster in the air. **ANSWER A**

G2D05 What is the most useful type of map to use when orienting a directional HF antenna toward a distant station?
    A. Azimuthal projection   C. Polar projection
    B. Mercator projection   D. Stereographic projection

An *azimuthal projection* map. (See page 78.) **ANSWER A**

G9D10 Which of the following describes a Beverage antenna?
    A. A vertical antenna constructed from beverage cans
    B. A broad-band mobile antenna
    C. A helical antenna for space reception
    D. A very long and low receiving antenna that is highly directional

If you have plenty of real estate, the *low and long beverage antenna* can achieve some great directionality. This lets you either home in on a specific signal, or minimize noise coming off a nearby power line. **ANSWER D**

G9D09 Which of the following is an application for a Beverage antenna?
A. Directional transmitting for low HF bands
B. Directional receiving for low HF bands
C. Portable Direction finding at higher HF frequencies
D. Portable Direction finding at lower HF frequencies

Some of the larger high frequency base station rigs have a separate antenna port for an optional *receive antenna*. This would be perfect for a *beverage antenna* that you can build yourself out of wire. **ANSWER B**

G9D08 Why is a Beverage antenna generally not used for transmitting?
A. Its impedance is too low for effective matching
B. It has high losses compared to other types of antennas
C. It has poor directivity
D. All of these choices are correct

Sounds like a soft drink, but it's actually a fabulous shortwave receive antenna – beverage. It is fabulous for pulling in distant stations with minimal atmospheric noise reception. However, use it as a receive only antenna, because its *transmit losses are comparatively higher* than the common dipole. **ANSWER B**

G4E05 Which of the following most limits the effectiveness of an HF mobile transceiver operating in the 75 meter band?
A. "Picket Fencing" signal variation
B. The wire gauge of the DC power line to the transceiver
C. The HF mobile antenna system
D. FCC rules limiting mobile output power on the 75 meter band

75 meters is a great nighttime band for talking more than 1,000 miles away. However, to get on 75 meters with a mobile unit, you need a monster antenna loading coil and quite possibly a capacity hat on your antenna system. Your *HF antenna system* will be your *biggest challenge* when *operating 75 meters, mobile*. **ANSWER C**

A Capacity Hat Mounted on 75 Meter Vertical Antenna Will Improve Mobile Performance.

G5A08 Why is impedance matching important?
A. So the source can deliver maximum power to the load
B. So the load will draw minimum power from the source
C. To ensure that there is less resistance than reactance in the circuit
D. To ensure that the resistance and reactance in the circuit are equal

When internal source and load *impedances are matched, maximum power* will be delivered to the load. Most new ham HF radios will automatically reduce power output when there is an impedance mismatch. **ANSWER A**

G5A14 Which of the following describes one method of impedance matching between two AC circuits?
A. Insert an LC network between the two circuits
B. Reduce the power output of the first circuit
C. Increase the power output of the first circuit
D. Insert a circulator between the two circuits
Two AC circuits might be *impedance matched* by using *coils and capacitors (LC)* between the two circuits. **ANSWER A**

G5A12 What is one reason to use an impedance matching transformer?
A. To reduce power dissipation in the transmitter
B. To maximize the transfer of power
C. To minimize SWR at the antenna
D. To minimize SWR in the transmission line
When *impedances are matched*, we will have the *greatest amount of power transfer*. An impedance matching transformer allows us to precisely match radio stages for the maximum transfer of power. **ANSWER B**

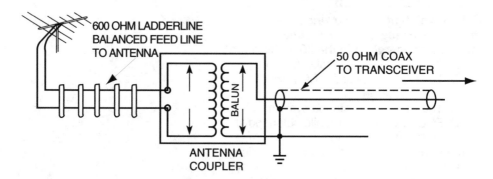

Use Antenna Coupler to Match Antenna Feedline to Coaxial Cable
Source: *Antennas – Selection and Installation*, © 1986 Master Publishing, Inc., Niles, IL

G5A13 Which of the following devices can be used for impedance matching at radio frequencies?
A. A transformer
B. A Pi-network
C. A length of transmission line
D. All of these choices are correct
There are plenty of ways we can match impedances at radio frequencies. Up at the antenna, we sometimes will use fractional wavelength impedance matching transmission lines. In a radio RF output stage, we might use an impedance matching Pi-network, and within the radio, small transformers will allow us to impedance match. *All of these* are great ways for providing the maximum transfer of radio frequency energy. **ANSWER D**

G5A07 What happens when the impedance of an electrical load is equal to the internal impedance of the power source?
A. The source delivers minimum power to the load
B. The electrical load is shorted
C. No current can flow through the circuit
D. The source can deliver maximum power to the load

Always make sure your new General Class worldwide antenna systems have an impedance around 50 ohms for *maximum power transfer*. Some General Class ham transceivers have built-in, automatic, *impedance-matching* antenna tuner networks. **ANSWER D**

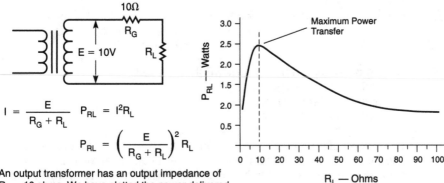

$$I = \frac{E}{R_G + R_L} \qquad P_{RL} = I^2 R_L$$

$$P_{RL} = \left(\frac{E}{R_G + R_L}\right)^2 R_L$$

An output transformer has an output impedance of R_G = 10 ohms. We have plotted the power delivered to R_L as R_L varies when E = 10V.
The values for the curve are calculated by substituting different values of R_L into the formula when E = 10V and $R_G = 10\Omega$.

Maximum power is transferred when $R_G = R_L$.

Maximum Power Transfer

G7A05 What should be the impedance of a low-pass filter as compared to the impedance of the transmission line into which it is inserted?
    A. Substantially higher   C. Substantially lower
    B. About the same   D. Twice the transmission line impedance
As we discussed earlier, *impedances should always be the same* for maximum transfer of power. Therefore, you want the low-pass filter to have the same impedance as both the transmission line and the ham transceiver to which it is connected. **ANSWER B**

G9A06 Which of the following is a reason for using an inductively coupled matching network between the transmitter and parallel conductor feed line feeding an antenna?
    A. To increase the radiation resistance
    B. To reduce spurious emissions
    C. To match the unbalanced transmitter output to the balanced parallel
       conductor feedline
    D. To reduce the feed-point impedance of the antenna
If you were to feed a center-fed folded dipole antenna with parallel conductor feed line, you would need a *matching network to match* the 50-ohm *unbalanced transmitter output to* the 300-ohm *balanced parallel conductor feed line*.
ANSWER C

G4B13 What is one measurement that can be made with a dip meter?

A. The resonant frequency of a circuit
B. The tilt of the ionosphere
C. The gain of an antenna
D. The notch depth of a filter

You can purchase a combination antenna analyzer that also doubles as a dip meter. If you need to *determine the resonant frequency* of some old antenna coils, or of a radio circuit, the *dip meter* will do the job. **ANSWER A**

G4B12 What is one way a noise bridge might be used?

A. Determining an antenna's gain in dBi
B. Pre-tuning an antenna tuner
C. Pre-tuning a linear amplifier
D. Determining the line loss of the antenna system

Many amateur radio operators still use manual antenna tuners. The practice is to first pre-tune the tuner on background noise in order to get it set relatively close before you transmit and twist the knobs for lowest SWR. But on 10 meters there might not be much noise, so the *noise bridge* is a great way for *pre-tuning your manual tuner*. But if it were me, I would get busy tuning the antenna, and trying to get rid of the need for an antenna tuner at all. **ANSWER B**

Antenna Noise Bridge
Courtesy of MFJ Enterprises, Inc.

G4B04 How is a noise bridge normally used?

A. It is connected at an antenna's feed point and reads the antenna's noise figure
B. It is connected between a transmitter and an antenna and tuned for minimum SWR
C. It is connected between a receiver and an antenna of unknown impedance and is adjusted for minimum noise
D. It is connected between an antenna and ground and tuned for minimum SWR

You may use a *noise bridge between* your amateur *receiver and an unknown impedance*. Tune the antenna noise bridge for minimum noise and then read the impedance on the dial. Never transmit when this device is connected in the line. **ANSWER C**

G4B14 Which of the following must be connected to an antenna analyzer when it is being used for SWR measurements?

A. Receiver
B. Transmitter
C. Antenna and feedline
D. All of these answers are correct

The portable *antenna analyzer* is hooked up *to the antenna and feedline* to determine a standing wave ratio *(SWR) measurement*. **ANSWER C**

G4B08 What instrument may be used to monitor relative RF output when making antenna and transmitter adjustments?

A. A field-strength meter
B. An antenna noise bridge
C. A multimeter
D. A Q meter

For relative transmitter output checks, an inexpensive *field-strength meter* is a good addition to any ham station. **ANSWER A**

G4B10 Which of the following can be determined with a field strength meter?

A. The radiation resistance of an antenna
B. The radiation pattern of an antenna
C. The presence and amount of phase distortion of a transmitter
D. The presence and amount of amplitude distortion of a transmitter

A well-calibrated *field-strength meter* might be used to *determine radiation field patterns* off that brand-new ground plane antenna you just built for 10 meters. But in the real world of radio, the only way to properly measure antenna pattern field strength is out in the middle of a 5-acre field where the ground is absolutely flat and there are no buildings or structures to reflect the signal. The antenna is then rotated with the field-strength measurements at a specific fixed spot. **ANSWER B**

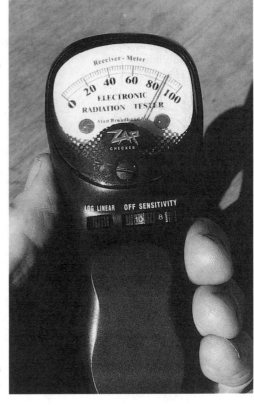

A hand-held field strength meter is a handy tool to have in your shack.

G4B11 Which of the following might be a use for a field strength meter?
A. Close-in radio direction-finding
B. A modulation monitor for a frequency or phase modulation transmitter
C. An overmodulation indicator for a SSB transmitter
D. A keying indicator for a RTTY or packet transmitter

Now here is a "great" correct answer that really describes the best use of a field-strength meter. A fun amateur radio sport is tracking down hidden milliwatt and microwatt transmitters, and the *field-strength meter* will really let you *"sniff" an area where* you think the *transmitter may be hidden*. T-hunt activities are now a worldwide sport! **ANSWER A**

Website Resources

▼ IF YOU'RE LOOKING FOR	▼ THEN VISIT
Roll Your Own Antenna · · · · · · · · · · · ·	www.cebik.com
Portable Satellite Beams · · · · · · · · · ·	www.arrowantennas.com
HF Antennas · · · · · · · · · · · · · · · · · ·	www.AORUSA.com
	www.directivesystems.com
	www.MFJEnterprises.com
	www.M2INC.com
	www.NEW-Tronics.com
	www.DiamondAntenna.com
	www.Cushcraft.com
	www.Cubex.com
	www.Radioworks.com

Coax Cable

G9A02 What is the typical characteristic impedance of coaxial cables used for antenna feedlines at amateur stations?

A. 25 and 30 ohms
B. 50 and 75 ohms
C. 80 and 100 ohms
D. 500 and 750 ohms

Amateur Radio coax cable usually is rated at *50 ohms* impedance. You might also use some very large TV hard-line coax cable with a proper matching network for your ham setup. Most CATV coax is rated at *75 ohms*. **ANSWER B**

This photo shows the different inside construction of various brands of coax cable

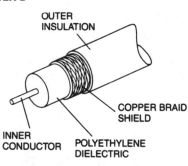

OUTER INSULATION

COPPER BRAID SHIELD

INNER CONDUCTOR POLYETHYLENE DIELECTRIC

Coaxial (Called Coax)

G9A07 How does the attenuation of coaxial cable change as the frequency of the signal it is carrying increases?

A. It is independent of frequency
B. It increases
C. It decreases
D. It reaches a maximum at approximately 18 MHz

The *higher* you go in *frequency*, the *greater the attenuation* of the transmission line. This is why it's very important to always use the largest size coax cable available for VHF and UHF frequencies. **ANSWER B**

G9A08 In what values are RF feedline losses usually expressed?

A. Ohms per 1000 ft
B. dB per 1000 ft
C. Ohms per 100 ft
D. dB per 100 ft

RF feedline losses are usually expressed in *decibels per 100 feet*. **ANSWER D**

Attenuation	
Frequency (MHz)	(dB/100 ft.)
2	0.21
10	0.5
20	0.71
100	1.7
200	2.4
1000	5.7

Attenuation of RG-8 Coax with Foam Dielectric

G5B13 What percentage of power loss would result from a transmission line loss of 1 dB?

A. 10.9 %
B. 12.2 %
C. 20.5 %
D. 25.9 %

On our examination, it will be expected that certain things are committed to memory within that gray matter. In the real world of ham radio, the memory cells would tell you to look at a chart to calculate decibel losses in a transmission line, or decibel gains in beam antenna effective radiated power.

In this question, they ask the percentage loss from a transmission line of 1 dB. *One dB of loss* results in .794 (79.4%) of the energy making it through the coax, leading to *20.6%* getting lost in the transmission line (100% - 79.4% = 20.6% loss, the correct answer). **ANSWER C**

G9A03 What is the characteristic impedance of flat ribbon TV type twin lead?

A. 50 ohms
B. 75 ohms
C. 100 ohms
D. 300 ohms

Amateur Radio coax cable usually is rated at 50 ohms impedance. You might also use some very large TV hard-line coax cable with a proper matching network for your ham setup. Most *twin lead is rated at 300 ohms.* **ANSWER D**

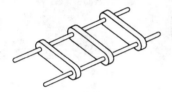

a. Parallel Two-Wire Line

b. Twisted Pair

c. Two-Wire Ribbon Flat Lead (Twin Lead)

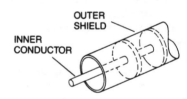

d. Air Coaxial with Washer Insulator

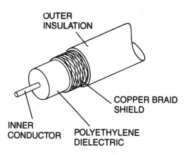

f. Coaxial (Called Coax)

Different Transmission Lines

Source: *Antennas – Selection and Installation,* © 1986 Master Publishing, Inc., Niles, IL

G9A01 Which of the following factors help determine the characteristic impedance of a parallel conductor antenna feedline?
    A. The distance between the centers of the conductors and the radius of the conductors
    B. The distance between the centers of the conductors and the length of the line
    C. The radius of the conductors and the frequency of the signal
    D. The frequency of the signal and the length of the line
It's important never to squash coax cable in a window or in a car door. If you do, it will change the *distance between the center conductor* and the outside braid, and this changes the characteristic impedance of the cable – whether it's coax or parallel conductor feed line. **ANSWER A**

G9A05 What must be done to prevent standing waves on an antenna feedline?
    A. The antenna feed point must be at DC ground potential
    B. The feedline must be cut to an odd number of electrical quarter wavelengths long
    C. The feedline must be cut to an even number of physical half wavelengths long
    D. The antenna feed point impedance must be matched to the characteristic impedance of the feedline
Always try to *match feedpoint impedance* of the antenna *to the characteristic impedance* of the feedline *for* maximum power transfer and *minimum SWR*. **ANSWER D**

G9A04 What is a common reason for the occurrence of reflected power at the point where a feedline connects to an antenna?
    A. Operating an antenna at its resonant frequency
    B. Using more transmitter power than the antenna can handle
    C. A difference between feedline impedance and antenna feed point impedance
    D. Feeding the antenna with unbalanced feedline
Keep the impedances of your feed line and antenna the same for minimum standing wave ratio. Standing waves are set-up on the feed line because *power* is being *reflected back from an impedance mismatch.* **ANSWER C**

G4B15 Which of the following can be measured with a directional wattmeter?
    A. Standing Wave Ratio   C. RF interference
    B. Antenna front-to-back ratio   D. Radio wave propagation
A *directional watt meter* will make a dandy *standing wave ratio* checker. Maximum forward power with minimum reflected power will indicate an acceptable low SWR. **ANSWER A**

G9A11 What standing-wave-ratio will result from the connection of a 50-ohm feed line to a non-reactive load having a 50-ohm impedance?
    A. 2:1   C. 50:50
    B. 1:1   D. 0:0
50 into 50 is a perfect match, so your SWR would be 1:1. **ANSWER B**

G9A12 What would be the SWR if you feed a vertical antenna that has a 25-ohm feed-point impedance with 50-ohm coaxial cable?
    A. 2:1
    B. 2.5:1
    C. 1.25:1
    D. You cannot determine SWR from impedance values

It's relatively easy to build a 10- or 15-meter ground plane antenna out of copper tubing. The radiating element is one-quarter wavelength long, and the ground radials are also one-quarter wavelength long. The ground radials must be bent down at an approximately 45-degree angle in order to bring the impedance up to 50 ohms, which is the normal impedance of coax cable. If the ground plane copper tubes extend straight out at 90 degrees from the base of the ground plane, the impedance might look like *25 ohms instead of 50 ohms*. This would result in an SWR reading of about *2:1*, acceptable, but easily improved by simply bending the ground radials down at a 45-degree angle. **ANSWER A**

G9A09 What standing-wave-ratio will result from the connection of a 50-ohm feed line to a non-reactive load having a 200-ohm impedance?
    A. 4:1   C. 2:1
    B. 1:4   D. 1:2

50 into 200 goes 4 times, so the impedance of the mismatch is *4:1*. 1:4 is not the correct answer, even though it looks correct. **ANSWER A**

Make sure your antenna and feedline impedances are matched
so your signal will be heard loud and clear around the world.

G9A10 What standing-wave-ratio will result from the connection of a 50-ohm feed line to a non-reactive load having a 10-ohm impedance?
    A. 2:1   C. 1:5
    B. 50:1   D. 5:1

10 into 50 goes 5 times, so the mismatch is *5:1*. 1:5 is not correct. **ANSWER D**

G9A14 If the SWR on an antenna feedline is 5 to 1, and a matching network at the transmitter end of the feedline is adjusted to 1 to 1 SWR, what is the resulting SWR on the feedline?

  A. 1:1

  B. 5:1

  C. Between 1:1 and 5:1 depending on the characteristic impedance of the line

  D. Between 1:1 and 5:1 depending on the reflected power at the transmitter

Grandpa is on the roof, fixing some shingles, and accidentally chops off a couple of feet at one end of your hidden dipole. The SWR shoots up to 5:1, and your new high frequency transceiver immediately pulls back power output. Most new transceivers with built-in automatic antenna tuners can't resolve SWR any higher than 3:1, so you drag out an old manual tuner, twiddle the knobs, and magically SWR now drops down to 1:1. Your radio is happy. Grandpa is not. He says every time your are on the air you blank out the TV. Guess what? Even though you reduced the high SWR to a "magic" 1:1 to the transceiver, you still have standing waves flowing back on the outside of the coax from the damaged antenna. This means the *feedline is still* sky high with and elevated *5:1 SWR*. Time to get back to the roof and fix the broken wire. **ANSWER B**

G9A13 What would be the SWR if you feed a folded dipole antenna that has a 300-ohm feed-point impedance with 50-ohm coaxial cable?

  A. 1.5:1

  B. 3:1

  C. 6:1

  D. You cannot determine SWR from impedance values

The folded dipole has a characteristic impedance of about 300 ohms. This is why it is many times fed with twin lead and the impedance transformed with the use of a manual antenna tuner. If you tried to hook up 50-ohm coax cable directly to a 300-ohm feedpoint, your *SWR would be* an unacceptable *6:1*. Divide 50 ohms into 300 ohms for the 6:1 ratio. **ANSWER C**

An automatic antenna tuner with built-in
SWR meter for non-resonant wire antennas.

G4D09 Which of these connector types is commonly used for RF service at frequencies up to 150 MHz?

  A. Octal   C. UHF

  B. RJ-11   D. DB-25

Your worldwide radio uses the common antenna jack called *"UHF"*. This jack will receive the common PL-259. On a high frequency radio that includes 2 meters and 440 MHz, and maybe even 1.2 GHz, you indeed might discover a separate Type N connector. **ANSWER C**

Elmer Point: *The common UHF jack on the back of your new HF transceiver will accept the coax cable PL-259 connector, which is the common coax connector for high frequency use. However, the PL-259 is NOT waterproof. If you use an antenna feed point that takes a PL-259, make absolutely sure to completely waterproof the connection with flexible sealant, as well as an added wrap of self-vulcanizing tape. The number one reason why ham radio antenna systems fail is water migrating into the PL-259 and killing the feed point connection. When I visit your QTH, I don't want to see any exposed PL-259 connectors on the coax at the antenna feed point!*

G4D07 Which of the following describes a Type-N connector?
A. A moisture resistant RF connector useful to 10 GHz
B. A small bayonet connector used for data circuits
C. A threaded connector used for hydraulic systems
D. An audio connector used in surround sound installations

As a licensed Technician Class operator, chances are you encountered the Type N connector when working with UHF equipment and microwave equipment. The *Type N connector is* great at VHF and UHF antenna feed points because it is *moisture resistant and it will work all the way up to 10 GHz*, which is 10,000 MHz. When working with your new high frequency SSB transceiver, it is doubtful that you will encounter a Type N connector. **ANSWER A**

BNC, Type N. and PL 259 Connectors

G1B01 What is the maximum height above ground to which an antenna structure may be erected without requiring notification to the FAA and registration with the FCC, provided it is not at or near a public-use airport?
A. 50 feet
B. 100 feet
C. 200 feet
D. 300 feet

Coax Cable

The FCC works closely with the FAA when it comes to towers taller than *200 feet*. Although you do not need FCC approval for towers under 200 feet, you may need approval from your city or homeowner's association. [97.15(a)] **ANSWER C**

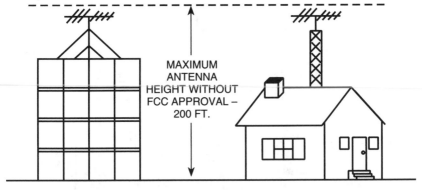

MAXIMUM ANTENNA HEIGHT WITHOUT FCC APPROVAL – 200 FT.

Maximum Antenna Height

Elmer Point: *When ham radio operators get together for an antenna party, or a club meeting for mobile and base antenna demonstration, one of the greatest safety hazards is the protruding element at eye level. Make absolutely sure all sharp antenna element ends have a soft, protective cover to prevent eye injury. If you are assembling an antenna and ready to put it up on a tower, make certain that everyone is wearing protective safety glasses and a hard hat. Eyesight is so precious and a sharp antenna element at eye level is downright dangerous. Always think eye safety when working around mobile and base antenna systems.*

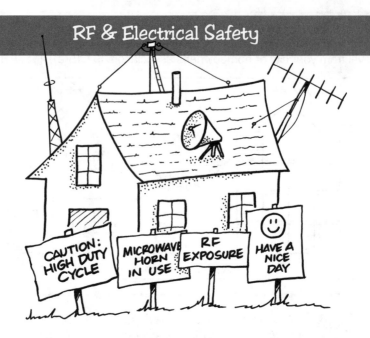

G0A12 What precaution should you take whenever you make adjustments or repairs to an antenna?

A. Ensure that you and the antenna structure are grounded
B. Turn off the transmitter and disconnect the feedline
C. Wear a radiation badge
D. All of these answers are correct

If you're going to be working on an antenna system, *disconnect the feedline from the transmitter* so there is absolutely no way that someone could accidentally transmit while you are aloft repairing the antenna. **ANSWER B**

Before going aloft to repair an antenna, always disconnect the feedline from the transmitter to prevent accidental exposure to RF radiation.

G0B08 What should be done by any person preparing to climb a tower that supports electrically powered devices?

A. Notify the electric company that a person will be working on the tower
B. Make sure all circuits that supply power to the tower are locked out and tagged
C. Ground the base of the tower
D. Disconnect the feedline for every antenna at the station

To insure that no one starts rotating a beam from the electric powered rotator on the top of the tower, double check that the rotator control unit down below is *unplugged and tagged "climber aloft."* **ANSWER B**

G0B07 Which of the following should be observed for safety when climbing on a tower using a safety belt or harness?

A. Never lean back and rely on the belt alone to support your weight
B. Always attach the belt safety hook to the belt "D" ring with the hook opening away from the tower
C. Ensure that all heavy tools are securely fastened to the belt D ring
D. Make sure that your belt is grounded at all times

Fall prevention from a tower is best achieved with a professional safety harness that regularly gets inspected, making sure the *belt's safety hook* attaches to the "D" ring with the *hook opening always away from the tower*. **ANSWER B**

G0A08 Which of the following steps must an amateur operator take to ensure compliance with RF safety regulations?

A. Post a copy of FCC Part 97 in the station
B. Post a copy of OET Bulletin 65 in the station
C. Perform a routine RF exposure evaluation
D. All of these choices are correct

Performing a routine RF exposure evaluation is a good idea for all amateurs to ensure compliance with RF-safety regulations, and to ensure that you and your neighbors are not becoming "overexposed" to RF radiation. **ANSWER C**

G0A11 What precaution should you take if you install an indoor transmitting antenna?

A. Locate the antenna close to your operating position to minimize feed line radiation
B. Position the antenna along the edge of a wall to reduce parasitic radiation
C. Make sure that MPE limits are not exceeded in occupied areas
D. No special precautions are necessary if SSB and CW are the only modes used

It's always a good idea to locate an antenna as far away as possible from living spaces that will be occupied when you are transmitting on the air. If you're just receiving, no problem – but *when you're transmitting*, to *minimize RF* exposure get your wire antenna away from everyone! **ANSWER C**

G0A01 What is one way that RF energy can affect human body tissue?

A. It heats body tissue
B. It causes radiation poisoning
C. It causes the blood count to reach a dangerously low level
D. It cools body tissue

General Class

When you reheat a slice of ham or beef in the microwave oven, what are you actually doing? The microwave oven concentrates radio signals into the meat and it heats up. If microwaves are concentrated on the human body, it *heats the body tissue* just like your leftovers. **ANSWER A**

Never stand in front of a microwave feedhorn antenna.
On transmit, it radiates a concentrated beam of RF energy.

G0A03 Which of the following has the most direct effect on the permitted exposure level of RF radiation?
A. The age of the person exposed
B. The power level and frequency of the energy
C. The environment near the transmitter
D. The type of transmission line used
The human body as a whole is resonant at frequencies from 30 MHz to 300 MHz. It is the *frequency*, or wavelength, *and* the *power level* of the radiofrequency energy has the most direct effect on our exposure level to RF radiation. **ANSWER B**

G0A04 What does "time averaging" mean in reference to RF radiation exposure?
A. The average time of day when the exposure occurs
B. The average time it takes RF radiation to have any long-term effect on the body
C. The total time of the exposure
D. The total RF exposure averaged over a certain time
Time-averaging is a method of calculating an individual's *total exposure to RF radiation over a given period of time*. The premise of time-averaging is that the human body can tolerate larger amounts of RF radiation if the exposure is received in short "bursts" as compared to a constant exposure at the same high level. As depicted here, total exposure to various levels of radiation is averaged over a 6-

minute period. On a time-averaged basis, the amount of thermal load on the body is equal in all cases. **ANSWER D**

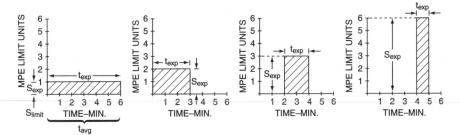

The General Equation for Time Averaging Exposure Equivalence is:

$$S_{exp}\ t_{exp}\ =\ S_{limit}\ t_{avg}$$

G0A07 What effect does transmitter duty cycle have when evaluating RF exposure?
   A. A lower transmitter duty cycle permits greater short-term exposure levels
   B. A higher transmitter duty cycle permits greater short-term exposure levels
   C. Low duty cycle transmitters are exempt from RF exposure evaluation requirements
   D. Only those transmitters that operate at a 100% duty cycle must be evaluated

Duty cycle is the percentage of time the transmitter is actually sending out energy. If you hold down your telegraph key for continuous full-transmit-power output, this would be 100% duty cycle. If the space in between each dit were equal to the duration of the transmitted signal, this would be a 50% duty cycle. The *lower duty cycle will permit greater short-term exposure* levels to RF radiation. **ANSWER A**

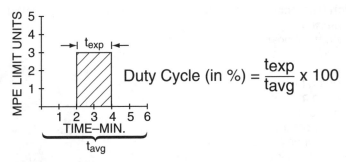

$$\text{Duty Cycle (in \%)} = \frac{t_{exp}}{t_{avg}} \times 100$$

Duty Cycle

G0A02 Which property is NOT important in estimating if an RF signal exceeds the maximum permissible exposure (MPE)?
   A. Its duty cycle   C. Its power density
   B. Its critical angle   D. Its frequency

The duty cycle of the incoming microwaves, their frequency, and their power density all determine how much radiofrequency energy is affecting the body tissue. The term *"critical angle" has nothing to do with RF damage* to the body. Critical angle deals with how radio waves are refracted by the ionosphere. **ANSWER B**

G0A15 How can you determine that your station complies with FCC RF exposure regulations?

A. By calculation based on FCC OET Bulletin 65
B. By calculation based on computer modeling
C. By measurement of field strength using calibrated equipment
D. All of these choices are correct

You can determine how your station complies with FCC RF exposure regulations by using The W5YI RF Safety Tables in the Appendix of this book. You can use the tables to estimate safe distances based on FCC OET Bulletin No. 65, or by your own calculations based on computer modeling. You also can actually go out there and measure with a field-strength meter the power density levels making sure to use calibrated equipment. *All of these choices* are a good way to determine whether or not you are going to expose yourself or your neighbors unnecessarily. [97.13(c)(1)] **ANSWER D**

G0A05 What must you do if an evaluation of your station shows RF energy radiated from your station exceeds permissible limits?

A. Take action to prevent human exposure to the excessive RF fields
B. File an Environmental Impact Statement (EIS-97) with the FCC
C. Secure written permission from your neighbors to operate above the controlled MPE limits
D. All of these answers are correct

Federal Communications Commission rules state that: "The licensee must perform the routine RF environmental evaluation prescribed by Section 1.1307(b) of this chapter, if the power of the licensee's station exceeds the following PEP limits." And before you do anything else, make sure you *take adequate precautions to prevent* you or your neighbors from being *overexposed to RF*. **ANSWER A**

G0A06 Which transmitter(s) at a multiple user site is/are responsible for RF safety compliance?

A. Only the most powerful transmitter on site
B. All transmitters on site, regardless of their power level or duty cycle
C. Any transmitter that contributes 5% or more of the MPE
D. Only those that operate at more than 50% duty cycle

Station owners operating repeaters need to do careful calculations to ensure that the addition of their transmitter won't cause the entire repeater site to reach overexposure. Each transmitter is included in the RF exposure site evaluation if it produces more than 5% of the maximum permissible power limit for that transmitter. *Remember "5%"* and notice that all the other answers are different. **ANSWER C**

All of the RF emissions from a repeater site like this one must be considered collectively when performing an RF safety evaluation.

G0A10 What do the RF safety rules require when the maximum power output capability of an otherwise compliant station is reduced?
 A. Filing of the changes with the FCC
 B. Recording of the power level changes in the log or station records
 C. Performance of a routine RF exposure evaluation
 D. No further action is required

One way to minimize your personal exposure to strong RF fields is to reduce power output. If you comply with the RF safety rules at 500 watts, you would also comply at a lesser power level, so *no further action* is required. **ANSWER D**

G0A09 What type of instrument can be used to accurately measure an RF field?
 A. A receiver with an S meter
 B. A calibrated field-strength meter with a calibrated antenna
 C. A betascope with a dummy antenna calibrated at 50 ohms
 D. An oscilloscope with a high-stability crystal marker generator

Although a simple field-strength meter can show the presence of radiofrequency emissions, it takes a precisely *CALIBRATED field-strength meter with a CALIBRATED antenna to accurately measure RF fields.* **ANSWER B**

G0A13 What precaution should be taken when installing a ground-mounted antenna?
 A. It should not be installed higher than you can reach
 B. It should not be installed in a wet area
 C. It should be painted so people or animals do not accidentally run into it
 D. It should be installed so no one can be exposed to RF radiation in excess of maximum permissible limits

There are all sorts of great 5-band and 7-band trap vertical antennas that work nicely on the ground, or on a metal shed just above the ground. When you're looking for a spot to mount the antenna, keep in mind maximum permissible exposure limits and *install it so that no one can actually walk up to it*, or stand near it when you are transmitting. This goes for your pets, too. Don't fry Fido or Furball by leaving a ground-mounted vertical unfenced. **ANSWER D**

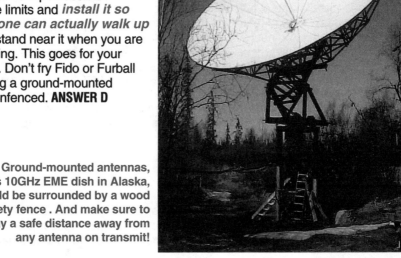

Ground-mounted antennas, like this 10GHz EME dish in Alaska, should be surrounded by a wood safety fence . And make sure to stay a safe distance away from any antenna on transmit!

G0A14 What is one thing that can be done if evaluation shows that a neighbor might receive more than the allowable limit of RF exposure from the main lobe of a directional antenna?
    A. Change from horizontal polarization to vertical polarization
    B. Change from horizontal polarization to circular polarization
    C. Use an antenna with a higher front-to-back ratio
    D. Take precautions to ensure that the antenna cannot be pointed at their house

If your calculations indicate you may be exposing your neighbors to too much RF, you may need to relocate your entire antenna system, or take precautions to *ensure it cannot be pointed at their house* when transmitting. **ANSWER D**

G4C10 Which of the following is covered in the National Electrical Code?
    A. Acceptable bandwidth limits
    B. Acceptable modulation limits
    C. Electrical safety inside the ham shack
    D. RF exposure limits of the human body

The National Electrical Code covers *electrical safety standards* as they relate to conductors. RF exposure limits to the human body are covered by ANSI. NEC covers electrical safety at your home station. **ANSWER C**

G0B05 Which of the following conditions will cause a Ground Fault Circuit Interrupter (GFCI) to disconnect the 120 or 240 Volt AC line power to a device?
    A. Current flowing from the hot wire to the neutral wire
    B. Current flowing from the hot wire to ground
    C. Over-voltage on the hot wire
    D. All of these choices are correct

Ground fault circuit interrupters are found in newer electrical sockets, and they will instantly open a circuit that detects *current flowing from the hot wire to ground*. **ANSWER B**

G0B06 Why must the metal chassis of every item of station equipment be grounded (assuming the item has such a chassis)?
    A. It prevents blowing of fuses in case of an internal short circuit
    B. It provides a ground reference for the internal circuitry
    C. It ensures that the neutral wire is grounded
    D. It ensures that hazardous voltages cannot appear on the chassis

Good station *grounding insures no* hazardous or *dangerous voltages* appearing on the metal chassis of your equipment. **ANSWER D**

G0B01 Which wire(s) in a four-conductor line cord should be attached to fuses or circuit breakers in a device operated from a 240-VAC single-phase source?
    A. Only the "hot" (black and red) wires
    B. Only the "neutral" (white) wire
    C. Only the ground (bare) wire
    D. All wires

Most ham transceivers run straight out of the box on 12 volts DC. The big huge honking high frequency base stations may have their own power supplies built-in – 110 VAC. However, kilowatt linear amplifiers may require 240 volts alternating current, and for single phase *240 VAC, we fuse the hot black and red wires*, and NEVER fuse the neutral white wire and NEVER fuse the neutral ground bare copper wire. Most amplifiers already have the fuse circuits built in. **ANSWER A**

G0B03 Which size of fuse or circuit breaker would be appropriate to use with a circuit that uses AWG number 14 wiring?

A. 100 amperes
B. 60 amperes
C. 30 amperes
D. 15 amperes

This one works out "F" for "F"-14 gauge wiring will handle *15 amps*. **ANSWER D**

G0B02 What is the minimum wire size that may be safely used for a circuit that draws up to 20 amperes of continuous current?

A. AWG number 20
B. AWG number 16
C. AWG number 12
D. AWG number 8

The minimum wire size to handle 20 amps should be *AWG #12*. Remember "T" for "T"-12, 20. **ANSWER C**

Wire Size A.W.G. (B&S)	Current-Amps (Continuous Duty)	
	Single Wire	Bundled Wire
8	73	46
10	55	33
12	41	23
14	32	17
16	22	13
18	16	10

American Wire Gauge (AWG) Wire Size vs. Current Capability

G0B12 What is the purpose of a transmitter power supply interlock?

A. To prevent unauthorized access to a transmitter
B. To guarantee that you cannot accidentally transmit out of band
C. To ensure that dangerous voltages are removed if the cabinet is opened
D. To shut off the transmitter if too much current is drawn

Linear amplifiers using vacuum tubes will likely have a power supply interlock switch. If the equipment is on and the *power supply door opened up, the equipment will* either instantly *power off*, or you will be greeted by a loud bang where the high voltage gets shorted to ground tripping an internal fuse or breaker. **ANSWER C**

G0B04 What is the mechanism by which electrical shock can be lethal?

A. Current through the heart can cause the heart to stop pumping
B. A large voltage field can induce currents in the brain
C. Heating effects in major organs can cause organ failure
D. All of these choices are correct

Current flowing in your hands and out your footsies could lead to a FATAL electrical shock. Any *current passing through the heart can cause the heart to stop pumping*. Never work on anything electrical in your bare feet! **ANSWER A**

G0B14 What is the maximum amount of electrical current flow through the human body that can be tolerated safely?
    A. 5 microamperes   C. 500 milliamperes
    B. 50 microamperes   D. 5 amperes
The maximum amount of electrical current that might flow through the body that can be tolerated safely is *ONLY 50 microamperes*. This is far less current than a typical nightlight might pull, so be careful around ANY electrical circuits where your body could accidentally become an inadvertent conductor. NEVER GOOD!
ANSWER B

G0B13 Which of the following is the most hazardous type of electrical energy?
    A. Direct Current
    B. 60 cycle Alternating current
    C. Radio Frequency
    D. All of these choices are correct
It has been proven that the most dangerous form of electrical energy is your common 60 cycle alternating current lines around the house or in the office. Watch out – *plain old 60 cycle AC can kill you!* Be safe – I want to see you upgrade to Extra Class soon. **ANSWER B**

G0B10 Which of the following is a danger from lead-tin solder?
    A. Lead can contaminate food if hands are not washed carefully after handling
    B. High voltages can cause lead-tin solder to disintegrate suddenly
    C. Tin in the solder can "cold flow" causing shorts in the circuit
    D. RF energy can convert the lead into a poisonous gas
Any time you are working on radio equipment innards, always *wash your hands after you get through with your great soldering job* to remove any lead that might be contained in the roll of old Granddad's solder. **ANSWER A**

G0B11 Which of the following is good engineering practice for lightning protection grounds?
    A. They must be bonded to all buried water and gas lines
    B. Bends in ground wires must be made as close as possible to a right angle
    C. Lightning grounds must be connected to all ungrounded wiring
    D. They must be bonded together with all other grounds
Good engineering practice for a lightening ground indicates *all grounds BONDED together with all other grounds*. **ANSWER D**

G0B09 Why is it not safe to use soldered joints with the wires that connect the base of a tower to a system of ground rods?
    A. The resistance of solder is too high
    B. Solder flux will prevent a low conductivity connection
    C. Solder has too high a dielectric constant to provide adequate lightning
        protection
    D. A soldered joint will likely be destroyed by the heat of a lightning strike
Carefully inspect tower grounding circuits where all ground connections have been swaged rather than soldered. *Solder might melt on a direct lightening strike.*
ANSWER D

G4E12 Which of the following is a primary reason for not placing a gasoline-fueled generator inside an occupied area?

A. Danger of carbon monoxide poisoning
B. Danger of engine over torque
C. Lack of oxygen for adequate combustion
D. Lack of nitrogen for adequate combustion

If you plan to operate from a ham radio emergency communications vehicle, make sure you spot the carbon monoxide alarm system. In rare cases, the running generator beneath the communications vehicle might leak *carbon monoxide* into the operating compartment. This could become LETHAL. Always think "fresh air" safety when operating in a vehicle with the engine or generator running. **ANSWER A**

G4E06 Which of the following is true of an emergency generator installation?

A. The generator should be located in a well ventilated area
B. The generator should be insulated from ground
C. Fuel should be stored near the generator for rapid refueling in case of an emergency
D. All of these choices are correct

When you set up for field day in June, chances are you'll be using an emergency generator to keep all of the stations on the air. *Make sure the generator is not located where some of the field day operators will inhale the toxic exhaust.* Make sure the generator is properly grounded, and always store the generator's fuel in a safe place away from any inhabited area. **ANSWER A**

Taking the General Class Examination

Get ready for worldwide skywave band privileges! As soon as you pass your General Class theory exam, you will be licensed to broadcast on frequency bands that regularly offer skywave excitement every hour of the day, and all through the night. This chapter tells you how the examination will be given, who is qualified to administer the General Class exam, and what happens after you successfully complete the written exam.

IMPORTANT NOTE: In order to receive credit for Technician Class, a prerequisite to the Element 3 General Class examination, you must bring a photocopy of your original, signed, valid Technician Class license to the examination site. You also will need 2 forms of identification, one of which must be a photo ID (such as your driver's license). If you can't find your Technician Class original license, you can download a copy from the FCC website at wireless.fcc.gov and then doing a license search by your call sign or your name. There no longer is any requirement to show a Morse code credit or to take a Morse code test.

If you are brand new to ham radio, you can certainly take both the Technician and General Class exams in one test session as long as you pass the Technician exam first.

THE GENERAL CLASS EXAMINATION

Here is an overview of the General Class examination and what to expect when you go to the test session.

Examination Administration

The General Class exam is given by a team of 3 Volunteer Examiners (VEs) – hams who hold Advanced or Extra Class licenses and who are accredited to administer your exam by a Volunteer Examiner Coordinator (VEC).

Volunteer Examiner Teams offer examinations on a regular basis at local sites to serve their communities. Generally, the VETs closely coordinate their activities with one another, so you should be able to find a nearby test site and exam date that is convenient for you. You can obtain information about VECs and exam sessions in your area by checking with your local radio club, ham radio store, or local packet bulletin boards. A list of VECs that was current at the time of publication is given in the Appendix (see page 200).

Once you have found your local VE team, contact them to select a test date and location and pre-register for your examination. They will hold a seat for you at the next available session. Don't be a no-show, and don't be a surprise-show. Call them ahead of time and pre-register!

The Volunteer Examiners are not compensated for their time and skills, but they are permitted to charge you a fee for certain reimbursable expenses incurred in preparing, administering, and processing the examination. The maximum fee is adjusted annually by the FCC and currently is about $14.00. When you call to make your exam reservation, ask the VE the current amount of the exam fee.

Want to find a test site fast?
Visit the W5YI-VEC website at: www.w5yi.org, or call them at 800-669-9594.

EXAM CONTENT

The questions, answers, and distracters for each question of the General Class written examination are public information. The question pool included in this book contains all 484 possible questions that can be used to make up your 35-question Element 3 written examination. The VEC is not permitted to change any of the wording, punctuation or numerical values included in any questions, answers, or distracters. The VET can change the A-B-C-D order of the answers, if it wishes.

Also, while the VET is not required to select one question from each syllabus topic under each subelement, because many of the questions within the same syllabus topic are duplicate questions asked in different ways, you should expect an exam that contains one question from each syllabus topic within each subelement. Look again at the question pool syllabus on page 209 to see how the test will be constructed.

WHAT TO BRING TO THE SITE

Here's what you'll need to bring with you for your General Class examination:
- The fee of approximately $14.00 in cash, exact change. No checks accepted.
- The original plus two copies of your current Technician or Technician-Plus license. If your license has not yet arrived, make sure you bring the original plus two copies of your Certificate of Successful Completion of Examination (CSCE) indicating your most current license status.
- A photo identification card – your driver's license is ideal for this purpose.
- Some sharp pencils and fine-tip pens. It's good to have a backup.
- Calculators may be used. However, the examiners may erase the memory before your exam begins.
- Any other items that the VET asks you to bring.

TAKING THE EXAM

Don't speed read the examination! Read each question carefully. Take your time looking for the correct answer. Some answers start out looking correct, but end up wrong. When you finish, go back over every question and double-check your answers. When you are satisfied that you have passed the examination, turn in all of your test papers to the examination team. Make sure to thank your VEs and to let them know how much you appreciate their efforts to help promote our hobby.

AFTER THE EXAM

Wait patiently outside the exam room for your results. Chances are the VEs will greet you with a smile and your CSCE. (If you didn't pass, they will tell you what to do next.) When you are told you passed the exam, be sure you are given the appropriate paperwork:

- The CSCE, signed by all three examiners.
- Make sure the temporary identifier is filled in so you can use it to immediately go on the air with your new General Class frequency privileges.

COMPLETING NCVEC FORM 605

When you arrive at the examination site, one of the first things you will do is complete the NCVEC Form 605. This form is retained by the Volunteer Exam Coordinator who transfers your printed information to an electronic file and sends it to the FCC for your new license, or upgrade. Your application may be delayed or kicked-back to you if the VEC can't read your writing. Make absolutely sure you print as legibly as you can, and carefully follow the instructions on the form.

NCVEC QUICK-FORM 605 APPLICATION FOR AMATEUR OPERATOR/PRIMARY STATION LICENSE

SECTION 1 - TO BE COMPLETED BY APPLICANT

PRINT LAST NAME	SUFFIX	FIRST NAME	INITIAL	STATION CALL SIGN (IF ANY)
MARCONI		JOE	G	KB9SMG

MAILING ADDRESS (Number and Street or P.O. Box)	SOCIAL SECURITY NUMBER / TIN (OR LICENSEE ID)
7101 RECTIFIER ROAD	090-909-090

CITY	STATE CODE	ZIP CODE (5 or 9 Numbers)	E-MAIL ADDRESS (OPTIONAL)
INDUCTOR	IL	60777	SPARKS@MSN.COM

DAYTIME TELEPHONE NUMBER (Include Area Code) OPTIONAL	FAX NUMBER (Include Area Code) OPTIONAL	ENTITY NAME (IF CLUB, MILITARY RECREATION, RACES)

Type of Applicant: [X] Individual [ ] Amateur Club [ ] Military Recreation [ ] RACES (Renewal Only) — TRUSTEE OR CUSTODIAN CALL SIGN

I HEREBY APPLY FOR (Make an X in the appropriate box(es)) — SIGNATURE OF RESPONSIBLE CLUB OFFICIAL

[ ] EXAMINATION for a **new** license grant   [ ] CHANGE my mailing address to **above** address
[X] EXAMINATION for **upgrade** of my license class   [ ] CHANGE my station **call sign** systematically
[ ] CHANGE my **name** on my license to my new name   Applicant's Initials: _____

NCVEC Form 605

Name

If you are upgrading from a current license, it is very important to compare the information on your present license with what you are writing on NCVEC Form 605. Make sure that *everything* on NCVEC Form 605 *is identical* to how your present license reads. Fill in your last name, first name, middle initial, and suffix such as junior or senior. You must stay absolutely consistent with your name on any future Form 605s for upgrades or changes of address. If you start out as "Jack" and end up "John," the computer will throw out your next application. If you decide to use a nickname, this is okay – but down the line when you visit a foreign country, they may ask you for identification that needs to illustrate this same nickname. It is best to stick with the name that is on most of your personal pictured IDs, such as your Driver's License.

Date of Birth

The biggest problem here is, without thinking, you put down this year's date rather than the year in which you were born. One out of twenty make this mistake.

Social Security Number & FRN

You are required to write in either your Social Security Number or your FCC Registration Number (FRN) in the designated box. If you are a citizen of another country, put down the country name in this box. Your current Technician Class license should show your FRN. If you do not have an FRN and you prefer not to disclose your Social Security Number, you should obtain an FRN *before* the exam session so that you can complete your Form 605 when you pass your test. To obtain an FCC Registration Number, go to the following website and follow the instructions there. Visit:

https://svartifoss2.fcc.gov/coresWeb/publicHome.do

Take our word for it – just give them your SSN and avoid a lot of grief.

Address

Where do you want your license mailed? To avoid contacting the FCC regularly for a change of address, use a mailing address that you plan to keep as permanent as possible. Has your address changed from the one on your current license? If so, check the "change" box and list your new address.

e-mail Address

This is optional, but it's a good idea because the amateur radio service is now under the FCC's Universal Licensing System. Once you get your new call sign, you will be able to work with the FCC directly via computer, including change of address, change of name, and license renewals without having to do any paperwork.

Phone Numbers

There are two boxes for phone numbers – one for a daytime contact, and the other for your FAX number. Put both numbers down in just in case the VEC or VE team need to re-contact you because they can't read your writing.

Signature

Sign your name as legibly as possible and include all of the letters that you printed as your name at the top of the form. Don't just put down a squiggle or an initial. You need to sign your name all the way out, including all of the letters that were in your printed name.

Final Check

Finally, double-check that your handwriting is legible. If a single letter in your name can't be read clearly and is misinterpreted, subsequent electronic filings may get returned as no action. Make sure your Form 605 is as clear as a bell to your Volunteer Examination team, who will then forward it to their VEC.

Your Examiners' Portion

The VET will carefully review your NCVEC Form 605 to ensure that they can read your handwriting and that everything looks okay. They will then enter this information into their computer database, and will most likely file your test passing results electronically to their Volunteer Examiner Coordinator. The VEC will then verify the information and electronically file your results with the FCC.

YOUR UPGRADE TO GENERAL CLASS

Usually, your General Class upgrade or new call sign will be granted within 72 hours of passing an examination, if it is electronically filed by the VEC – and you should receive a paper copy of your new General Class license in about 3 weeks. When you pass your written exam, you will be issued a CSCE – Certificate of Successful Completion of Examination. This CSCE allows you to begin using your new privileges *immediately*. After your call sign, append the letters "AG" to indicate your upgrade is being processed by the FCC. As soon as you see your upgrade on the electronic database, or receive your new license, you can drop the "AG" at the end of your call sign. Visit the W5YI-VEC website, where you will find links to other sites that allow you to look-up your upgrade or new call sign. The address is www.w5yi.org.

   If you are going from no license to General Class, *you may not operate immediately* with your CSCE because you have no call sign. But thanks to electronic filing, your new call sign should show up on the FCC database in about 3 to 5 days.

GENERAL CLASS CALL SIGNS

The FCC has exhausted the availability of "Group C" Technician and General Class call signs that begin with the letter "N," a number, and three other letters, such as N9ABC.

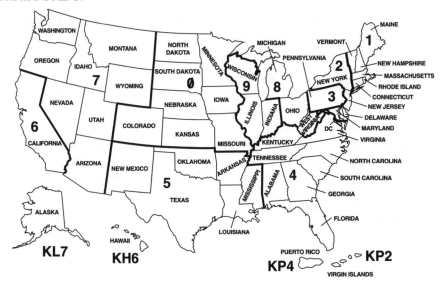

U.S. Call Sign Areas

If you did not check the "Change Call Sign" box on your NCVEC Form 605 application, you will simply keep your current call sign. However, if you do check the "Change Call Sign" box, you will receive an entry-level "Group D" call sign as if you were a newly-licensed amateur. I suggest you don't check the "Change Call Sign" box and stick with your present call sign.

Vanity Call Signs

You are eligible to replace your computer-generated, no-choice call sign with a vanity call sign of your choosing. This call sign could be made up of your initials, or represent your love of animals (K9DOG) or could be call letters that your late mom or dad had when they got started in ham radio years ago. General Class amateur operators may request a vanity call sign from Group D or Group C. You also may request a call sign that was previously assigned to you that may have expired years ago, as well as a call sign of a close relative or former holder who is now deceased. This call sign can be from any group.

There is an additional fee for a vanity call sign. You must complete FCC Form 605 and Form 159, and attach your check payable to the Federal Communications Commission to Form 159. Both forms are available on the FCC internet site: **www.fcc.gov/wtb/amateur**. You can file Form 605 electronically, and then mail in the Form 159 with your check attached. Electronically filed forms for which the filing fee and Form 159 have been received will be processed first. Instructions for electronic filing are included on the FCC website. Mail the required paperwork and fee to: FCC, PO Box 358994, Pittsburgh, PA 15251-5994. The Form 159 and the fee must be received within 10 days of electronically filing your Form 605 or your application will be dismissed.

To make it easier for you to select a vanity call sign, I suggest you contact the W5YI Group at 800-669-9594 and ask for their vanity call sign application package. The W5YI Group can help you to file electronically so you end up with exact call sign of your choice.

As for me, I'm staying with my original-issue WB6NOA call sign. If I changed it, I would be breaking a 50-year tradition!

CONGRATULATIONS! YOU PASSED!

After you pass the written exam and code test, congratulations are in order and we offer you a big welcome to the worldwide privileges of General Class! Day or night, summer or fall, sunshine or rain, there is always a worldwide band open and ready for General Class voice, code, PACTOR, and television communications. Your new worldwide privileges are added to your existing VHF and UHF privileges. Before you go on the air, do a lot of listening. This will assure that you get started on the right foot with your new privileges. Remember, even the worldwide bands have band plans, so make sure you are operating within the plan for your worldwide communications.

I also would like you to write me so I can send you an exclusive General Class passing certificate, plus some valuable manufacturers' discount coupons. Send a self-addressed large envelope with 12 first class stamps loose on the inside to: Gordon West, WB6NOA, Radio School, Inc., 2414 College Drive, Costa Mesa, CA 92626.

Once again... *Welcome to General Class!* I hope to work you on one of the worldwide bands very soon. You can catch me mobile in our communications van regularly on 14.240 MHz, and I'm a "regular" on 10 meters from my Southern California home at 28.400 MHz.

Your next upgrade is *Extra Class*. For Extra Class, use my 3rd Ham Book, *Extra Class*. My explanations make the formula problem-solving easy and the learning fun! So start thinking about that upgrade now.

It's been fun teaching you the General Class. Good luck on that upcoming exam – I know you're going to pass!

73!

Gordon West, WB6NOA

PS: Keep reading! Chapter 5 will get you started on your journey to learn CW!
    Aw, come on – you can do it!!

5

Learning Morse Code

On April 15, 2000, the FCC dropped the 20- and 13-word-per-minute Morse code requirements for worldwide frequency privileges for General and Extra Class operators down to 5 words per minute. On February 23, 2007, the FCC *totally eliminated* the Morse code test as a prerequisite for high frequency operation.

The elimination of the Morse code test for operation on worldwide frequencies conforms to international radio regulations. In 2003, the International Telecommunications Union (ITU) World Radiocommunication Conference voted to allow individual nations to determine whether or not to retain a Morse code test as a requirement to operate on frequencies below 30 MHz.

When the FCC eliminated the Morse code test for Technician Class operators in 1991 for VHF/UHF operating, the ruling was adopted with little opposition. However, the announcement that the FCC was considering total elimination of the Morse code test drew thousands of written comments to the FCC. Many comments supported code test elimination, while a minority urged the FCC to retain a code test because of the strong tradition of CW as a ham radio operating mode.

The Federal Communications Commission concluded that "...this change (eliminating the code test) eliminates an unnecessary burden that may discourage current amateur radio operators from advancing their skills and participating more fully in the benefits of amateur radio." The FCC Commissioners recognized the simple fact that learning Morse code was keeping many very technical, talented hams from obtaining their General and Extra Class licenses. Morse code is much like musical rhythms. Some people are tone deaf, and some people couldn't carry a rhythm in a hand basket.

So for years, the Morse code test was an insurmountable hurdle to many talented Technician Class hams who wanted to upgrade. If I could take these Techs and put them into one of my regular Morse code classes, we usually could get the majority of them through the CW test with outside home study, on the air practice (on 2 meters), and classroom study followed by the code test. But throughout the country, Morse code classes were few and far between and it is tough to learn new music and a new language without classroom instruction.

Technician Class operators could not practice on the worldwide airwaves to learn the code because these bands were reserved for only those operators who had already passed the code test. Running Morse code practice on a local 2 meter repeater was one option, but nothing beats the excitement of practicing code on the worldwide bands and hooking up with another station thousands of miles away.

As of February 23, 2007, we can now take new General Class operators and introduce them to Morse code on the exciting worldwide bands!

Morse code is the ham radio operator's most basic language of short and long sounds, dits and dahs, or dots and dashes. Sailors have pounded SOS when trapped beneath a sailboat hull. In submarines, the tapping of Morse code gets the message through when there is no other way to communicate. Prisoners of war have tapped out Morse code messages, or BLINKED the code when being publicly displayed on television.

Ham operators use the code to get through when noise would otherwise cover up data or voice signals. Years ago, before road rage, fellow hams driving might greet each other by sending on their car horns H-I, a friendly salute to another ham.

So I encourage you to learn code. It is best mastered by sound along with memorizing the Morse code patterns seen on the upcoming pages. The pages show the number of dots and dashes to learn for a specific character, and learning the sound (rhythm) of Morse code is always the best way to practice. Let's see what these short sounds and long sounds are all about.

LOOKING AT MORSE CODE

The International Morse code, originally developed as the American Morse code by Samuel Morse, is truly international — all countries use it, and most commercial worldwide services employ operators who can recognize it. It is made up of short and long duration sounds. Long sounds, called "dahs," are three times longer than short sounds, called "dits." *Figure 2-1* shows the time intervals for Morse code sounds and spaces. *Figure 2-3,* on the next page, indicates the sounds for all the CW characters and symbols.

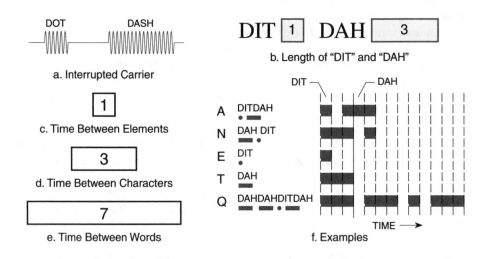

Figure 2-1. Time Intervals for Morse Code

Learning Morse Code

a. Alphabet

LETTER	Composed of:	Sounds like:	LETTER	Composed of:	Sounds like:
A	■ —	didah	N	— ■	dahdit
B	— ■ ■ ■	dahdididit	O	— — —	dahdahdah
C	— ■ — ■	dahdidahdit	P	■ — — ■	didahdahdit
D	— ■ ■	dahdidit	Q	— — ■ —	dahdahdidah
E	■	dit	R	■ — ■	didahdit
F	■ ■ — ■	dididahdit	S	■ ■ ■	dididit
G	— — ■	dahdahdit	T	—	dah
H	■ ■ ■ ■	didididit	U	■ ■ —	dididah
I	■ ■	didit	V	■ ■ ■ —	didididah
J	■ — — —	didahdahdah	W	■ — —	ditdahdah
K	— ■ —	dahdidah	X	— ■ ■ —	dahdididah
L	■ — ■ ■	didahdidit	Y	— ■ — —	dahdidahdah
M	— —	dahdah	Z	— — ■ ■	dahdahdidit

b. Special Signals and Punctuation

CHARACTER	Meaning:	Composed of:	Sounds like:
A̅R̅	(end of message)	■ — ■ — ■	didahdidahdit
K	invitation to transmit (go ahead)	— ■ —	dahdidah
S̅K̅	End of work	■ ■ ■ — ■ —	dididdidahdidah
S̅O̅S̅	International distress call	■ ■ ■ — — — ■ ■ ■	didididahdahdahdididit
V	Test letter (V)	■ ■ ■ —	didididah
R	Received, OK	■ — ■	didahdit
B̅T̅	Break or Pause	— ■ ■ ■ —	dahdidididah
D̅N̅	Slant Bar	— ■ ■ — ■	dahdidididahdit
K̅N̅	Back to You Only	— ■ — — ■	dahdidahdahdit
Period		■ — ■ — ■ —	didahdidahdidah
Comma		— — ■ ■ — —	dahdahdididahdah
Question mark		■ ■ — — ■ ■	dididahdahdidit
@	For Web Address	■ — — ■ — ■	didahdahdidahdi

c. Numerals

NUMBER	Composed of:	Sounds like:
1	■ — — — —	didahdahdahdah
2	■ ■ — — —	dididahdahdah
3	■ ■ ■ — —	didididahdah
4	■ ■ ■ ■ —	dididididah
5	■ ■ ■ ■ ■	didididit
6	— ■ ■ ■ ■	dahdidididit
7	— — ■ ■ ■	dahdahdididit
8	— — — ■ ■	dahdahdahdidit
9	— — — — ■	dahdahdahdahdit
Ø	— — — — —	dahdahdahdahdah

Figure 2-3. Morse Code and Its Sound

CODE KEY

Morse code is usually sent by using a code key. A typical one is shown in
Figure 2-2a. Normally it is mounted on a thin piece of wood or plexiglass. Make
sure that what you mount it on is thin; if the key is raised too high, it will be
uncomfortable to the wrist. The correct sending position for the hand is shown in
Figure 2-2b.

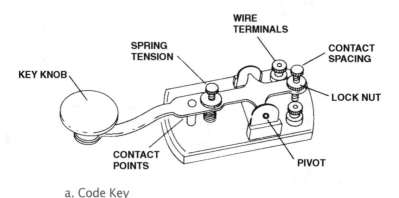

a. Code Key

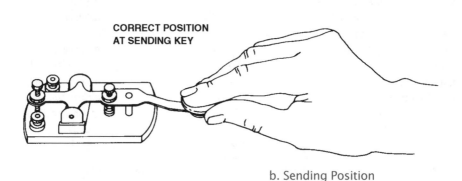

b. Sending Position

Figure 2-2. Code Key for Sending Code

LEARNING MORSE CODE

The reason you are learning the Morse code is to be able to operate all modes on the worldwide bands—including CW. Here are five suggestions (four serious ones) on how to learn the code:

1. Memorize the code from the code charts in this book.
2. Use my fun audio course available at all ham radio stores, and from the W5YI Group.
3. Go out and spend $1,000 and buy a worldwide radio, and listen to the code live and on the air. You don't need to spend that much, but you can listen to Morse code practice on the air, as shown in *Table 2-1.*
4. Use a code key and oscillator to practice sending the code. Believe it or not, someday you're actually going to do code over the live airwaves, using this same code key hooked up to your new megabuck transceiver.
5. Play with code programs on your computer, and *have fun!*

Table 2-1. Radio Frequencies and Times for Code Reception

Pacific	Mountain	Central	Eastern	Mon.	Tue.	Wed.	Thu.	Fri.
6 a.m.	7 a.m.	8 a.m.	9 a.m.		Fast Code	Slow Code	Fast Code	Slow Code
7 a.m. – 1 p.m.	8 a.m. – 2 p.m.	9 a.m. – 3 p.m.	10 a.m. – 4 p.m.	**VISITING OPERATOR TIME**				
1 p.m.	2 p.m.	3 p.m.	4 p.m.	Fast Code	Slow Code	Fast Code	Slow Code	Fast Code
2 p.m.	3 p.m.	4 p.m.	5 p.m.	Code Bulletin				
3 p.m.	4 p.m.	5 p.m.	6 p.m.	Digital Bulletin				
4 p.m.	5 p.m.	6 p.m.	7 p.m.	Slow Code	Fast Code	Slow Code	Fast Code	Slow Code
5 p.m.	6 p.m.	7 p.m.	8 p.m.	Code Bulletin				
6 p.m.	7 p.m.	8 p.m.	9 p.m.	Digital Bulletin				
6:45 p.m.	7:45 p.m.	8:45 p.m.	9:45 p.m.	Voice Bulletin				
7 p.m.	8 p.m.	9 p.m.	10 p.m.	Fast Code	Slow Code	Fast Code	Slow Code	Fast Code
8 p.m.	9 p.m.	10 p.m.	11 p.m.	Code Bulletin				

CW is broadcast on the following MHz frequencies: 1.8175, 3.5815, 7.0475, 14.0475, 18.0975, 21.0675, 28.0675, and 147.555. W1AW schedule courtesy of *QST* magazine.

CODE COURSES ON CDs AND CASSETTE TAPES

Five words per minute is so slow, and so easy, that many ham radio applicants learn it completely in a single week! You can do it, too, by using the code CDs and tapes mentioned above.

Code courses personally recorded by me make code learning *fun*. They will train you to send and receive the International Morse code in just a few short weeks. They are narrated and parallel the instructions in this book. The CDs have code characters generated at a 15-wpm character rate, spaced out to a 5-wpm word rate. This is known as Farnsworth spacing.

Getting Started

The hardest part of learning the code is taking the first CD out of the case, putting it in your player, and pushing the play button! Try it, and you will be over your biggest hurdle. After that, the CDs will talk you through the code in no time at all.

The CDs make code learning *fun*. You'll hear how humor has been added to the learning process to keep your interest high. Since ham radio is a hobby, there's no reason we can't poke ourselves in the ribs and have a little fun learning the code as part of this hobby experience. Okay, you're still not convinced — you probably have already made up your mind that trying to learn the code will be the hardest part of being a ham. It will not. Give yourself a fair chance. Don't get discouraged. Have patience and remember these important reminders when practicing to learn the Morse code:

- Learn the code by sound. Don't stare at the tiny dots and dashes that we have here in the book — the dit and dah sounds on the CDs and on the air and with your practice keyer will ultimately create an instant letter at your fingertips and into the pencil.
- *Never* scribble down dots or dashes if you forget a letter. Just put a small dash on your paper for a missed letter. You can go back and figure out what the word is by the letters you did copy!
- Practice only with fast code characters; 15-wpm character speed, spaced down to 5-wpm speed, is ideal.
- Practice the code by writing it down whenever possible. This further trains your brain and hand to work together in a subconscious response to the sounds you hear. (Remember Pavlov and his dog "Spot"?)
- Practice only for 15 minutes at a time. The CDs will tell you when to start and when to stop. Your brain and hand will lose that sharp edge once you go beyond 16 minutes of continuous code copy. You will learn much faster with five 15-minute practices per day than a one-hour marathon at night.
- Stay on course with the cassette instructions. Learn the letters, numbers, punctuation marks, and operating signals in the order they are presented. My code teaching system parallels that of the American Radio Relay League, Boy Scouts of America, the Armed Forces, and has worked for thousands in actual classroom instruction.

It was no accident that Samuel Morse gave the single dit for the letter "E" which occurs most often in the English language. He determined the most used letters in the alphabet by counting letters in a printer's type case. He reasoned a printer would have more of the most commonly-used letters. It worked! With just the first lesson, you will be creating simple words and simple sentences with no previous background.

Table 2-2 shows the sequence of letters, punctuation marks, operating signals, and numbers covered in six lessons on the CDs recorded by me.

Table 2-2. Sequence of Lessons on Cassettes

• Lesson 1	E T M A N I S O $\overline{SK}$ Period
• Lesson 2	R U D C 5 Ø $\overline{AR}$ Question Mark
• Lesson 3	K P B G W F H $\overline{BT}$ Comma
• Lesson 4	Q L Y J X V Z $\overline{DN}$ 1 2 3 4 6 7 8 9
• Lesson 5	Random code with narrated answers
• Lesson 6	A typical 5-wpm code test

CODE KEY AND OSCILLATOR – HAM RECEIVER

All worldwide ham transceivers have provisions for a code key to be plugged in for both CW practice off the air as well as CW operating on the air. If you already own a worldwide set, chances are all you will need is a code key for some additional code-sending practice.

Read over your worldwide radio instruction manual where it talks about hooking up the code key. For code practice, read the notes about operating with a "side tone" but not actually going on the air. This "side tone" capability of most worldwide radios will eliminate your need for a separate code oscillator.

Code Key and Oscillator — Separate Unit

Many students may wish to simply buy a complete code key and oscillator set. They are available from local electronic outlets or through advertisements in the ham magazines.

Look again at the code key in *Figure 2-2a*. Note the terminals for the wires. Connect wires to these terminals and tighten the terminals so the wires won't come loose. The two wires will go either to a code oscillator set or to a plug that connects into your ham transceiver. Hook up the wires to the plug as described in your ham transceiver instruction book or the code oscillator set instruction book.

Mount the key firmly, as previously described, then adjust the gap between the contact points. With most new telegraph keys, you will need a pair of pliers to loosen the contact adjustment knob. It's located on the very end of your keyer. First loosen the lock nut, then screw down the adjustment until you get a gap no wider than the thickness of a business card. You want as little space as possible between the points. The contact points are located close to the sending plastic knob.

Now turn on your set or oscillator and listen. If your hear a constant tone, check that the right-hand movable shorting bar is not closed. If it is, swing it open. Adjust the spring tension adjustment screw so that you get a good "feel" each time you push down on the key knob. Adjust it tight enough to keep the contacts from closing while your fingers are resting on the key knob.

Pick up the key by the knob! This is the exact position your fingers should grasp the knob—one or two on top, and one or two on the side of it. Poking at the knob with one finger is unacceptable. Letting your fingers fly off the knob between dots and dashes (dits and dahs) also is not correct. As you are sending, you should be able to instantly pick up the whole key assembly to verify proper finger position.

Your arm and wrist should barely move as you send CW. All the action is in your hand — and it should be almost effortless. Give it a try, and look at *Figure 2-2b* again to double-check your hand position.

Letting someone else use the key to send CW to you will also help you learn the code.

Morse Code Computer Software

The newest way to learn Morse code is through computer-aided instruction. There are many good PC programs on the market that not only teach you the characters, but build speed and allow you to take actual telegraphy examinations, which the computer constructs. Personal computer programs also can be used to make audio tapes on your tape recorder so you can listen to them on the cassette player in your car.

A big advantage of computer-aided Morse code learning is that you can easily customize the program to fit your own needs! You can select the sending speed, Farnsworth character-spacing speed, duration of transmission, number of characters in a random group, tone frequency — and more!

Some have built-in "weighting." That means the software will determine your weaknesses and automatically adjust future sending to give you more study on your problem characters! All Morse code software programs transmit the tone by keying the PC's internal speaker. Some generate a clearer audio tone through the use of external oscillators or internal computer sound cards.

Get those 6 code CDs and start listening to my voice, and see how easy it is to master the dots and dashes. Continuously push yourself to the letters with more dots and dashes in them, and work those tapes regularly and keep your copy in your spiral-bound notebook.

I hope to hear your CW call on the worldwide bands soon!

Need Gordo's Code CDs or CW software?
Call the W5YI Group at 1-800-669-9594, or visit www.w5yi.org

APPENDIX

U.S. VOLUNTEER EXAMINER COORDINATORS IN THE AMATEUR SERVICE

Anchorage Amateur Radio Club
P.O. Box 112573
Anchorage, AK 99511-2573
907/338-0662
e-mail: jwiley@alaska.net

ARRL/VEC
225 Main Street
Newington, CT 06111-1494
860/594-0300
860/594-0339 (fax)
e-mail: vec@arrl.org
Internet: www.arrl.org

Central America VEC, Inc.
1215 Dale Drive SE
Huntsville, AL 35801-2031
256/536-3904
256/534-5557 (fax)
e-mail: dontunstill@hamfest.org

Golden Empire Amateur Radio Society
P.O. Box 508
Chico, CA 95927-0508
530/345-3515
e-mail: wa6zrt@sbcglobal.net

Greater Los Angeles Amateur Radio Group
9737 Noble Avenue
North Hills, CA 91343-2403
818/892-2068
e-mail: gla.arg@gte.net
Internet: www.glaarg.org

Jefferson Amateur Radio Club
9912 Paula Dr.
River Ridge, LA 70123-1920
e-mail: doug@bellsouth.net

Laurel Amateur Radio Club, Inc.
P.O. Box 1259
Laurel, MD 20725-1259
301/937-0394 (6-9 PM)
e-mail: aa3of@arrl.net
Internet: http://larcmdorg.doore.net/vec/

The Milwaukee Radio Amateurs Club, Inc.
P.O. Box 070695
Milwaukee, WI 53207-0695
262/797-6722
e-mail: tom@supremecom.biz
Internet: www.w9rh.org

MO-KAN/VEC
228 Tennessee Road
Richmond, KS 66080-9174
785/867-2011
Internet: wo0e@pgtv.net

SANDARC-VEC
P.O. Box 2446
La Mesa, CA 91943-2446
619/697-1475
e-mail: n6nyx@cox.net
Internet: www.sandarc.net

Sunnyvale VEC Amateur Radio Club, Inc.
P.O. Box 60307
Sunnyvale, CA 94088-0307
408/255-9000 (exam info 24 hours)
e-mail: vec@amateur-radio.org
Internet: www.amateur-radio.org

W4VEC
P.O. Box 41
Lexington, NC 27293-0041
336/841-7576
e-mail: raef@lexcominc.net
Internet: www.w4vec.com

Western Carolina Amateur Radio Society/
    VEC, Inc.
6702 Matterhorn Ct.
Knoxville, TN 37918-6314
865/687-5410
e-mail: wcars@discoveret.org
Internet: www.discoveret.org/wcars

W5YI-VEC
P.O. Box 565101
Dallas, TX 75356-5101
817/860-3800
800-669-9495
e-mail: w5yi-vec@w5yi.org
Internet: www.w5yi.org

Appendix

THE W5YI RF SAFETY TABLES

(Developed by Fred Maia, W5YI, working in cooperation with the ARRL.)

There are two ways to determine whether your station's radio frequency signal radiation is within the MPE (Maximum Permissible Exposure) guidelines established by the FCC for *"controlled"* and *"uncontrolled"* environments. One way is direct *"measurement"* of the RF fields. The second way is through *"prediction"* using various antenna modeling, equations and calculation methods described in the FCC's *OET Bulletin 65* and *Supplement B*.

In general, most amateurs will not have access to the appropriate calibrated equipment to make precise field strength/power density measurements. The field-strength meters in common use by amateur operators are inexpensive, hand-held field strength meters that do not provide the accuracy necessary for reliable measurements, especially when different frequencies may be encountered at a given measurement location. It is more practical for amateurs to determine their PEP output power at the antenna and then look up the required distances to the controlled/uncontrolled environments using the following tables, which were developed using the prediction equations supplied by the FCC.

The FCC has determined that radio operators and their families are in the "controlled" environment and your neighbors and passers-by are in the "uncontrolled" environment. The estimated minimum compliance distances are in meters from the transmitting antenna to either the occupational/controlled exposure environment ("Con") or the general population/uncontrolled exposure environment ("Unc") using typical antenna gains for the amateur service and assuming 100% duty cycle and maximum surface reflection. Therefore, these charts represent the worst case scenario. They do not take into consideration compliance distance reductions that would be caused by:

(1) Feed line losses, which reduce power output at the antenna especially at the VHF and higher frequency levels.

(2) Duty cycle caused by the emission type. The emission type factor accounts for the fact that, for some modulated emission types that have a non-constant envelope, the PEP can be considerably larger than the average power. Multiply the distances by 0.4 if you are using CW Morse telegraphy, and by 0.2 for two-way SSB (single sideband) voice. There is no reduction for FM.

(3) Duty cycle caused by on/off time or "time-averaging." The RF safety guidelines permit RF exposures to be averaged over certain periods of time with the average not to exceed the limit for continuous exposure. The averaging time for occupational/controlled exposures is 6 minutes, while the averaging time for general population/uncontrolled exposures is 30 minutes. For example, if the relevant time interval for time-averaging is 6 minutes, an amateur could be exposed to two times the applicable power density limit for three minutes as long as he or she were not exposed at all for the preceding or following three minutes.

A routine evaluation is not required for vehicular mobile or hand-held transceiver stations. Amateur Radio operators should be aware, however, of the potential for exposure to RF electromagnetic fields from these stations, and take measures (such as reducing transmitting power to the minimum necessary, positioning the radiating antenna as far from humans as practical, and limiting continuous transmitting time) to protect themselves and the occupants of their vehicles.

Amateur Radio operators should also be aware that the FCC radio-frequency safety regulations address exposure to people — and not the strength of the signal. Amateurs may exceed the Maximum Permissible Exposure (MPE) limits as long as no one is exposed to the radiation.

How to read the chart: If you are radiating 500 watts from your 10 meter dipole (about a 3 dB gain), there must be at least 4.5 meters (about 15 feet) between you (and your family) and the antenna — and a distance of 10 meters (about 33 feet) between the antenna and your neighbors.

Medium and High Frequency Amateur Bands
All distances are in meters

Freq. (MF/HF) (MHz/Band)	Antenna Gain (dBi)	Peak Envelope Power (watts)							
		100 watts		500 watts		1000 watts		1500 watts	
		Con.	Unc.	Con.	Unc.	Con.	Unc.	Con.	Unc.
2.0 (160m)	0	0.1	0.2	0.3	0.5	0.5	0.7	0.6	0.8
2.0 (160m)	3	0.2	0.3	0.5	0.7	0.6	1.06	0.8	1.2
4.0 (75/80m)	0	0.2	0.4	0.4	1.0	0.6	1.3	0.7	1.6
4.0 (75/80m)	3	0.3	0.6	0.6	1.3	0.9	1.9	1.0	2.3
7.3 (40m)	0	0.3	0.8	0.8	1.7	1.1	2.5	1.3	3.0
7.3 (40m)	3	0.5	1.1	1.1	2.5	1.6	3.5	1.9	4.2
7.3 (40m)	6	0.7	1.5	1.5	3.5	2.2	4.9	2.7	6.0
10.15 (30m)	0	0.5	1.1	1.1	2.4	1.5	3.4	1.9	4.2
10.15 (30m)	3	0.7	1.5	1.5	3.4	2.2	4.8	2.6	5.9
10.15 (30m)	6	1.0	2.2	2.2	4.8	3.0	6.8	3.7	8.3
14.35 (20m)	0	0.7	1.5	1.5	3.4	2.2	4.8	2.6	5.9
14.35 (20m)	3	1.0	2.2	2.2	4.8	3.0	6.8	3.7	8.4
14.35 (20m)	6	1.4	3.0	3.0	6.8	4.3	9.6	5.3	11.8
14.35 (20m)	9	1.9	4.3	4.3	9.6	6.1	13.6	7.5	16.7
18.168 (17m)	0	0.9	1.9	1.9	4.3	2.7	6.1	3.3	7.5
18.168 (17m)	3	1.2	2.7	2.7	6.1	3.9	8.6	4.7	10.6
18.168 (17m)	6	1.7	3.9	3.9	8.6	5.5	12.2	6.7	14.9
18.168 (17m)	9	2.4	5.4	5.4	12.2	7.7	17.2	9.4	21.1
21.145 (15m)	0	1.0	2.3	2.3	5.1	3.2	7.2	4.0	8.8
21.145 (15m)	3	1.4	3.2	3.2	7.2	4.6	10.2	5.6	12.5
21.145 (15m)	6	2.0	4.6	4.6	10.2	6.4	14.4	7.9	17.6
21.145 (15m)	9	2.9	6.4	6.4	14.4	9.1	20.3	11.1	24.9
24.99 (12m)	0	1.2	2.7	2.7	5.9	3.8	8.4	4.6	10.3
24.99 (12m)	3	1.7	3.8	3.8	8.4	5.3	11.9	6.5	14.5
24.99 (12m)	6	2.4	5.3	5.3	11.9	7.5	16.8	9.2	20.5
24.99 (12m)	9	3.4	7.5	7.5	16.8	10.6	23.7	13.0	29.0
29.7 (10m)	0	1.4	3.2	3.2	7.1	4.5	10.0	5.5	12.2
29.7 (10m)	3	2.0	4.5	4.5	10.0	6.3	14.1	7.7	17.3
29.7 (10m)	6	2.8	6.3	6.3	14.1	8.9	19.9	10.9	24.4
29.7 (10m)	9	4.0	8.9	8.9	19.9	12.6	28.2	15.4	34.5

VHF/UHF Amateur Bands

All distances are in meters

Freq. (MF/HF) (MHz/Band)	Antenna Gain (dBi)	Peak Envelope Power (watts)							
		50 watts		100 watts		500 watts		1000 watts	
		Con.	Unc.	Con.	Unc.	Con.	Unc.	Con.	Unc.
50 (6m)	0	1.0	2.3	1.4	3.2	3.2	7.1	4.5	10.1
50 (6m)	3	1.4	3.2	2.0	4.5	4.5	10.1	6.4	14.3
50 (6m)	6	2.0	4.5	2.8	6.4	6.4	14.2	9.0	20.1
50 (6m)	9	2.8	6.4	4.0	9.0	9.0	20.1	12.7	28.4
50 (6m)	12	4.0	9.0	5.7	12.7	12.7	28.4	18.0	40.2
50 (6m)	15	5.7	12.7	8.0	18.0	18.0	40.2	25.4	56.8
144 (2m)	0	1.0	2.3	1.4	3.2	3.2	7.1	4.5	10.1
144 (2m)	3	1.4	3.2	2.0	4.5	4.5	10.1	6.4	14.3
144 (2m)	6	2.0	4.5	2.8	6.4	6.4	14.2	9.0	20.1
144 (2m)	9	2.8	6.4	4.0	9.0	9.0	20.1	12.7	28.4
144 (2m)	12	4.0	9.0	5.7	12.7	12.7	28.4	18.0	40.2
144 (2m)	15	5.7	12.7	8.0	18.0	18.0	40.2	25.4	56.8
144 (2m)	20	10.1	22.6	14.3	32.0	32.0	71.4	45.1	101.0
222 (1.25m)	0	1.0	2.3	1.4	3.2	3.2	7.1	4.5	10.1
222 (1.25m)	3	1.4	3.2	2.0	4.5	4.5	10.1	6.4	14.3
222 (1.25m)	6	2.0	4.5	2.8	6.4	6.4	14.2	9.0	20.1
222 (1.25m)	9	2.8	6.4	4.0	9.0	9.0	20.1	12.7	28.4
222 (1.25m)	12	4.0	9.0	5.7	12.7	12.7	28.4	18.0	40.2
222 (1.25m)	15	5.7	12.7	8.0	18.0	18.0	40.2	25.4	56.8
450 (70cm)	0	0.8	1.8	1.2	2.6	2.6	5.8	3.7	8.2
450 (70cm)	3	1.2	2.6	1.6	3.7	3.7	8.2	5.2	11.6
450 (70cm)	6	1.6	3.7	2.3	5.2	5.2	11.6	7.4	16.4
450 (70cm)	9	2.3	5.2	3.3	7.3	7.3	16.4	10.4	23.2
450 (70cm)	12	3.3	7.3	4.6	10.4	10.4	23.2	14.7	32.8
902 (33cm)	0	0.6	1.3	0.8	1.8	1.8	4.1	2.6	5.8
902 (33cm)	3	0.8	1.8	1.2	2.6	2.6	5.8	3.7	8.2
902 (33cm)	6	1.2	2.6	1.6	3.7	3.7	8.2	5.2	11.6
902 (33cm)	9	1.6	3.7	2.3	5.2	5.2	11.6	7.3	16.4
902 (33cm)	12	2.3	5.2	3.3	7.3	7.3	16.4	10.4	23.2
1240 (23cm)	0	0.5	1.1	0.7	1.6	1.6	3.5	2.2	5.0
1240 (23cm)	3	0.7	1.6	1.0	2.2	2.2	5.0	3.1	7.0
1240 (23cm)	6	1.0	2.2	1.4	3.1	3.1	7.0	4.4	9.9
1240 (23cm)	9	1.4	3.1	2.0	4.4	4.4	9.9	6.3	14.0
1240 (23cm)	12	2.0	4.4	2.8	6.2	6.2	14.0	8.8	19.8

All distances are in meters. To convert from meters to feet multiply meters by 3.28. Distance indicated is shortest line-of-sight distance to point where MPE limit for appropriate exposure tier is predicted to occur.

AUTHORIZED FREQUENCY BANDS – AMATEUR SERVICE (for U.S. Amateur Stations operating from ITU-Region 2–North and South America)

Current License Class[1] METERS	Grandfathered[2] Novice	Technician	General	Advanced	Extra Class
160			1800-2000 kHz/All	1800-2000 kHz/All	1800-2000 kHz/All
80 / 75	3525-3600 kHz/CW	3525-3600 kHz/CW	3525-3600 kHz/CW 3800-4000 kHz/Ph	3525-3600 kHz/CW 3700-4000 kHz/Ph	3500-4000 kHz/CW 3600-4000 kHz/Ph
40	7025-7125 kHz/CW	7025-7125 kHz/CW	7025-7125 kHz/CW 7175-7300 kHz/Ph	7025-7125 kHz/CW 7125-7300 kHz/Ph	7000-7300 kHz/CW 7125-7300 kHz/Ph
30			10.1-10.15 MHz/CW	10.1-10.15 MHz/CW	10.1-10.15 MHz/CW
20			14.025-14.15 MHz/CW 14.225-14.35 MHz/Ph	14.025-14.15 MHz/CW 14.175-14.35 MHz/Ph	14.0-14.35 MHz/CW 14.15-14.35 MHz/Ph
17			18.068-18.11 MHz/CW 18.11-18.168 MHz/Ph	18.068-18.11 MHz/CW 18.11-18.168 MHz/Ph	18.068-18.11 MHz/CW 18.11-18.168 MHz/Ph
15	21.025-21.2 MHz/CW	21.025-21.2 MHz/CW	21.025-21.2 MHz/CW 21.275-21.45 MHz/Ph	21.025-21.2 MHz/CW 21.225-21.45 MHz/Ph	21.0-21.45 MHz/CW 21.2-21.45 MHz/Ph
12			24.89-24.99 MHz/CW 24.93-24.99 MHz/Ph	24.89-24.99 MHz/CW 24.93-24.99 MHz/Ph	24.89-24.99 MHz/CW 24.93-24.99 MHz/Ph
10	28.0-28.5 MHz/CW 28.3-28.5 MHz/Ph	28.0-28.5 MHz/CW 28.3-28.5 MHz/Ph	28.0-28.3 MHz/CW 28.3-29.7 MHz/Ph	28.0-28.3 MHz/CW 28.3-29.7 MHz/Ph	28.0-29.7 MHz/CW 28.3-29.7 MHz/Ph
6		50-54 MHz/CW 50.1-54 MHz/Ph	50-54 MHz/CW 50.1-54 MHz/Ph	50-54 MHz/CW 50.1-54 MHz/Ph	50-54 MHz/CW 50.1-54 MHz/Ph
2		144-148 MHz/CW 144.1-148 MHz/All	144-148 MHz/CW 144.1-148 MHz/Ph	144-148 MHz/CW 144.1-148 MHz/All	144-148 MHz/CW 144.1-148 MHz/All
1.25		222-225 MHz/All [3]	222-225 MHz/All	222-225 MHz/All	222-225 MHz/All
0.70		420-450 MHz/All	420-450 MHz/All	420-450 MHz/All	420-450 MHz/All
0.33		902-928 MHz/All	902-928 MHz/All	902-928 MHz/All	902-928 MHz/All
0.23	1270-1295 MHz/All	1240-1300 MHz/All	1240-1300 MHz/All	1240-1300 MHz/All	1240-1300 MHz/All

[1] Effective 4-15-00   [2] Prior to 4-15-00   [3] Effective 2/1/94 219-220 MHz is authorized for point-to-point fixed digital message forwarding systems.

Note: Morse code (CW, A1A) may be used on any frequency allocated to the amateur service. Telephony emission (abbreviated Ph above) authorized on certain bands as indicated. Higher class licensees may use slow-scan television and facsimile emissions on the Phone bands; radio teletype/digital on the CW bands. All amateur modes and emissions are authorized above 144.1 MHz. In actual practice, the modes/emissions used are somewhat more complicated than shown above due to the existence of various band plans and "gentlemen's agreements" concerning where certain operations should take place.

Appendix

The following CEPT countries allow U.S. Amateurs to operate in their countries without a reciprocal license. Be sure to carry a copy of your FCC license and FCC Public Notice DA99-1098.

Austria	Finland	Liechtenstein	Slovenia
Belgium	France & its	Lithuania	Spain
Bosnia & Herzegovina	possessions	Luxembourg	Sweden
Bulgaria	Germany	Monaco	Switzerland
Croatia	Greenland	Netherlands	Turkey
Cyprus	Hungary	Netherlands Antilles	United Kingdom & its
Czech Republic	Iceland	Norway	possessions
Denmark	Ireland	Portugal	
Estonia	Italy	Romania	
Faroe Islands	Latvia	Slovak Republic	

List of Countries Permitting Third-Party Traffic

Country	Call Sign Prefix	Country	Call Sign Prefix	Country	Call Sign Prefix
Antigua and Barbuda	V2	El Salvador	YS	Paraguay	ZP
Argentina	LU	The Gambia	C5	Peru	OA
Australia	VK	Ghana	9G	Philippines	DU
Austria, Vienna	4U1VIC	Grenada	J3	St. Christopher & Nevis	V4
Belize	V3	Guatemala	TG	St. Lucia	J6
Bolivia	CP	Guyana	8R	St. Vincent & Grenadines	J8
Bosnia-Herzegovina	T9	Haiti	HH	Sierra Leone	9L
Brazil	PY	Honduras	HR	South Africa	ZS
Canada	VE, VO, VY	Israel	4X	Swaziland	3D6
Chile	CE	Jamaica	6Y	Trinidad and Tobago	9Y
Colombia	HK	Jordan	JY	Turkey	TA
Comoros	D6	Liberia	EL	United Kingdom	GB*
Costa Rica	TI	Marshall Is	V6	Uruguay	CX
Cuba	CO	Mexico	XE	Venezuela	YV
Dominica	J7	Micronesia	V6	ITU-Geneva	4U1ITU
Dominican Republic	HI	Nicaragua	YN	VIC-Vienna	4U1VIC
Ecuador	HC	Panama	HP		

Countries Holding U.S. Reciprocal Agreements

Antigua, Barbuda	Chile	Greece	Liberia	Seychelles
Argentina	Colombia	Greenland	Luxembourg	Sierra Leone
Australia	Costa Rica	Grenada	Macedonia	Solomon Islands
Austria	Croatia	Guatemala	Marshall Is.	South Africa
Bahamas	Cyprus	Guyana	Mexico	Spain
Barbados	Denmark	Haiti	Micronesia	St. Lucia
Belgium	Dominica	Honduras	Monaco	St. Vincent and
Belize	Dominican Rep.	Iceland	Netherlands	Grenadines
Bolivia	Ecuador	India	Netherlands Ant.	Surinam
Bosnia-	El Salvador	Indonesia	New Zealand	Sweden
Herzegovina	Fiji	Ireland	Nicaragua	Switzerland
Botswana	Finland	Israel	Norway	Thailand
Brazil	France[2]	Italy	Panama	Trinidad, Tobago
Canada[1]	Germany	Jamaica	Paraguay	Turkey
		Japan	Papua New Guinea	Tuvalu
		Jordan	Peru	United Kingdom[3]
		Kiribati	Philippines	Uruguay
		Kuwait	Portugal	Venezuela

1. Do not need reciprocal permit
2. Includes all French Territories
3. Includes all British Territories

POPULAR Q SIGNALS

Given below are a number of Q signals whose meanings most often need to be expressed with brevity and clarity in amateur work. (Q abbreviations take the form of questions only when each is sent followed by a question mark.)

QRG Will you tell me my exact frequency (or that of _____)? Your exact frequency (or that of _____) is _____ kHz.

QRH Does my frequency vary? Your frequency varies.

QRI How is the tone of my transmission? The tone of your transmission is _____ (1. Good; 2. Variable; 3. Bad).

QRJ Are you receiving me badly? I cannot receive you. Your signals are too weak.

QRK What is the intelligibility of my signals (or those of _____)? The intelligibility of your signals (or those of _____) is _____ (1. Bad; 2. Poor; 3. Fair; 4. Good; 5. Excellent).

QRL Are you busy? I am busy (or I am busy with _____). Please do not interfere.

QRM Is my transmission being interfered with? Your transmission is being interfered with _____ (1. Nil; 2. Slightly; 3. Moderately; 4. Severely; 5. Extremely).

QRN Are you troubled by static? I am troubled by static _____ (1-5 as under QRM).

QRO Shall I increase power? Increase power.

QRP Shall I decrease power? Decrease power.

QRQ Shall I send faster? Send faster (_____ WPM).

QRS Shall I send more slowly? Send more slowly (_____ WPM).

QRT Shall I stop sending? Stop sending.

QRU Have you anything for me? I have nothing for you.

QRV Are you ready? I am ready.

QRW Shall I inform _____ that you are calling on _____ kHz? Please inform _____ that I am calling on _____ kHz.

QRX When will you call me again? I will call you again at _____ hours (on _____ kHz).

QRY What is my turn? Your turn is numbered _____ .

QRZ Who is calling me? You are being called by _____ (on _____ kHz).

QSA What is the strength of my signals (or those of _____)? The strength of your signals (or those of _____) is _____ (1. Scarcely perceptible; 2. Weak; 3. Fairly good; 4. Good; 5. Very good).

QSB Are my signals fading? Your signals are fading.

QSD Is my keying defective? Your keying is defective.

QSG Shall I send _____ messages at a time? Send _____ messages at a time.

QSK Can you hear me between your signals and if so can I break in on your transmission? I can hear you between my signals; break in on my transmission.

QSL Can you acknowledge receipt? I am acknowledging receipt.

QSM Shall I repeat the last message which I sent you, or some previous message? Repeat the last message which you sent me [or message(s) number(s) _____].

QSN Did you hear me (or _____) on _____ kHz? I heard you (or _____) on _____ kHz.

QSO Can you communicate with _____ direct or by relay? I can communicate with _____ direct (or by relay through _____).

QSP Will you relay to _____ ? I will relay to _____ .

QST General call preceding a message addressed to all amateurs and ARRL members. This is in effect "CQ ARRL."

QSU Shall I send or reply on this frequency (or on _____ kHz)?

QSW Will you send on this frequency (or on _____ kHz)? I am going to send on this frequency (or on _____ kHz).

QSX Will you listen to _____ on _____ kHz? I am listening to _____ on _____ kHz.

QSY Shall I change to transmission on another frequency? Change to transmission on another frequency (or on _____ kHz).

QSZ Shall I send each word or group more than once? Send each word or group twice (or _____ times).

QTA Shall I cancel message number _____ ? Cancel message number _____ .

QTB Do you agree with my counting of words? I do not agree with your counting of words. I will repeat the first letter or digit of each word or group.

QTC How many messages have you to send? I have messages for you (or for _____).

QTH What is your location? My location is _____ .

QTR What is the correct time? The time is _____ .

Source: ARRL

COMMON CW ABBREVIATIONS

AA	All after	NW	Now; I resume transmission
AB	All before	OB	Old boy
ABT	About	OM	Old man
ADR	Address	OP-OPR	Operator
AGN	Again	OT	Old timer; old top
ANT	Antenna	PBL	Preable
BCI	Broadcast interference	PSE-PLS	Please
BK	Break; break me; break in	PWR	Power
BN	All between; been	PX	Press
B4	Before	R	Received as transmitted; are
C	Yes	RCD	Received
CFM	Confirm; I confirm	REF	Refer to; referring to; reference
CK	Check	RPT	Repeat; I repeat
CL	I am closing my station; call	SED	Said
CLD-CLG	Called; calling	SEZ	Says
CUD	Could	SIG	Signature; signal
CUL	See you later	SKED	Schedule
CUM	Come	SRI	Sorry
CW	Continuous Wave	SVC	Service; prefix to service message
DLD-DLVD	Delivered	TFC	Traffic
DX	Distance	TMW	Tomorrow
FB	Fine business; excellent	TNX	Thanks
GA	Go ahead (or resume sending)	TU	Thank you
GB	Good-by	TVI	Television interference
GBA	Give better address	TXT	Text
GE	Good evening	UR-URS	Your; you're; yours
GG	Going	VFO-	Variable-frequency oscillator
GM	Good morning	VY	Very
GN	Good night	WA	Word after
GND	Ground	WB	Word before
GUD	Good	WD-WDS	Word; words
HI	The telegraphic laugh; high	WKD-WKG	Worked; working
HR	Here; hear	WL	Well; will
HV	Have	WUD	Would
HW	How	WX	Weather
LID	A poor operator	XMTR	Transmitter
MILS	Milliamperes	XTAL	Crystal
MSG	Message; prefix to radiogram	XYL	Wife
N	No	YL	Young lady
ND	Nothing doing	73	Best regards
NIL	Nothing; I have nothing for you	88	Love and kisses
NR	Number		

SCHEMATIC SYMBOLS

RESISTORS	CAPACITORS	SWITCHES			BATTERIES		IC AMPLIFIERS	LOGIC (U#)	
FIXED	FIXED	SPST	SPDT	NORMAL OPEN	SINGLE CELL	MULTI CELL	GENERAL AMPLIFIER	AND	NAND
ADJUSTABLE	VARIABLE	TOGGLE		NORMAL CLOSED	GROUNDS		OP AMP	OR	NOR
WIRING		MULTPOINT		MOMENTARY	CHASSIS	EARTH		XOR	INVERT
CONDUCTORS NOT JOINED	CONDUCTORS JOINED								

INDUCTORS		TRANSISTORS			RELAYS		
FIXED	VARIABLE	NPN	P-CHANNEL	P-CHANNEL	SPST	SPDT	DPDT
DIODES	**TRANSFORMERS**	B C E	G D S	G D S			
LED (DS#)	AIR CORE						
DIODE/ RECTIFIER		PNP	N-CHANNEL	N-CHANNEL			
SCHOTTKY	WITH CORE	B C E	G D S	G D S	**ANTENNA**	**SPEAKER CRYSTAL**	
		BIPOLAR	SINGLE-GATE DEPLETION MODE MOSFET	SINGLE-GATE ENHANCEMENT MODE MOSFET	Y OR Y	OR	

SCIENTIFIC NOTATION

Prefix	Symbol		Multiplication Factor
exa	E	10^{18} =	1,000,000,000,000,000,000
peta	P	10^{15} =	1,000,000,000,000,000
tera	T	10^{12} =	1,000,000,000,000
giga	G	10^{9} =	1,000,000,000
mega	M	10^{6} =	1,000,000
kilo	k	10^{3} =	1,000
hecto	h	10^{2} =	100
deca	da	10^{1} =	10
(unit)		10^{0} =	1

Prefix	Symbol		Multiplication Factor
deci	d	10^{-1} =	0.1
centi	c	10^{-2} =	0.01
milli	m	10^{-3} =	0.001
micro	μ	10^{-6} =	0.000001
nano	n	10^{-9} =	0.000000001
pico	p	10^{-12} =	0.000000000001
femto	f	10^{-15} =	0.000000000000001
atto	a	10^{-18} =	0.000000000000000001

Appendix

QUESTION POOL SYLLABUS

The syllabus used by the NCVEC Question Pool Committee to develop the question pool is included here as an aid in studying the subelements and topic groups. Reviewing the syllabus will give you an understanding of how the question pool is used to develop the Element 3 General Class written examination. Remember, one question will be taken from each topic group within each subelement to create your exam.

2007-11 Element 3 General Class Syllabus

G1 – Commission's Rules
[5 Exam Questions - 5 Groups]
G1A General class control operator frequency privileges; primary and secondary allocations
G1B Antenna structure limitations; good engineering and good amateur practice; beacon operation; restricted operation; retransmitting radio signals
G1C Transmitter power regulations; HF data emission standards
G1D Volunteer Examiners and Volunteer Examiner Coordinators; temporary identification
G1E Control categories; repeater regulations; harmful interference; third party rules; ITU regions

G2 – Operating Procedures
[6 Exam Questions - 6 Groups]
G2A Phone operating procedures; USB/LSB utilization conventions; procedural signals; breaking into a QSO in progress; VOX operation
G2B Operating courtesy; band plans
G2C Emergencies, including drills and emergency communications
G2D Amateur auxiliary; minimizing interference; HF operations
G2E Digital operating: procedures, procedural signals and common abbreviations
G2F CW operating procedures and procedural signals, Q signals and common abbreviations; full break in

G3 – Radio Wave Propagation
[3 Exam Questions - 3 Groups]
G3A Sunspots and solar radiation; ionospheric disturbances; propagation forecasting and indices
G3B Maximum Usable Frequency; Lowest Usable Frequency; propagation "hops"
G3C Ionospheric layers; critical angle and frequency; HF scatter; Near Vertical Incidence Sky waves

G4 – Amateur Radio Practices
[5 Questions - 5 groups]
G4A Two-tone test; amplifier tuning and neutralization; DSP
G4B Test and monitoring equipment
G4C Interference with consumer electronics; grounding
G4D Speech processors; S meters; common connectors
G4E HF mobile radio installations; emergency and battery powered operation

G5 – Electrical Principles
[3 exam questions – 3 groups]
G5A Resistance; reactance; inductance; capacitance; impedance; impedance matching
G5B The Decibel; current and voltage dividers; electrical power calculations; sine wave root-mean-square (RMS) values; PEP calculations
G5C Resistors, capacitors, and inductors in series and parallel; transformers

G6 – Circuit Components
[3 exam question – 3 groups]
G6A - Resistors; capacitors; inductors
G6B - Rectifiers; solid state diodes and transistors; solar cells; vacuum tubes; batteries
G6C - Analog and digital integrated circuits (IC's); microprocessors; memory; I/O devices; microwave IC's (MMIC's); display devices

G7 – Practical Circuits
[2 exam question – 2 groups]
G7A Power supplies; transmitters and receivers; filters; schematic symbols
G7B Digital circuits (gates, flip-flops, shift registers); amplifiers and oscillators

G8 – Signals & Emissions
[2 exam questions – 2 groups]
G8A Carriers and modulation: AM; FM; single and double sideband ; modulation envelope; deviation; overmodulation
G8B Frequency mixing; multiplication; HF data communications; bandwidths of various modes

G9 – Antennas
[4 exam questions – 4 groups]
G9A Antenna feedlines: characteristic impedance, and attenuation; SWR calculation, measurement and effects; matching networks
G9B Basic antennas
G9C Directional antennas
G9D Specialized antennas

G0 – Electrical & RF Safety
[2 Exam Questions – 2 groups]
G0A RF safety principles, rules and guidelines; routine station evaluation
G0B Safety in the ham shack: electrical shock and treatment, grounding, fusing, interlocks, wiring, antenna and tower safety

2007-11 ELEMENT 3 Q&A CROSS REFERENCE

The following cross reference presents all 486 question numbers in numerical order included in the 2007-11 Element 3 General Class question pool, followed by the page number on which the question begins in the book. This will allow you to locate specific questions by question number. Note: two questions were deleted from the pool by the Question Pool Committee, resulting in an active pool of 484 questions. The deleted questions do not appear in the book.

Question	Page	Question	Page	Question	Page	Question	Page	Question	Page
G1 – Commission's		G1D08	23	G2C07	68	**G3 – Radio Wave**		**G4 – Amateur Radio**	
Rules		G1D09	23	G2C08	66	**Propagation**		**Practices**	
G1A01	25	G1D10	43	G2C09	66	G3A01	84	G4A01	103
G1A02	28	G1D11	42	G2C10	66	G3A02	82	G4A02	90
G1A03	28	G1D12	42	G2C11	67	G3A03	82	G4A03	12
G1A04	28	G1D13	41	G2C12	67	G3A04	80	G4A04	102
G1A05	28					G3A05	81	G4A05	103
G1A06	26	G1E01	39	G2D01	34	G3A06	82	G4A06	99
G1A07	31	G1E02	25	G2D02	35	G3A07	82	G4A07	99
G1A08	27	G1E03	47	G2D03	35	G3A08	83	G4A08	99
G1A09	31	G1E04	47	G2D04	78	G3A09	80	G4A09	99
G1A10	27	G1E05	39	G2D05	158	G3A10	80	G4A10	92
G1A11	26	G1E06	48	G2D06	79	G3A11	79	G4A11	90
G1A12	31	G1E07	40	G2D07	30	G3A12	81	G4A12	91
G1A13	32	G1E08	40	G2D08	34	G3A13	82	G4A13	103
G1A14	30	G1E09	24	G2D09	34	G3A14	84		
G1A15	30	G1E10	40	G2D10	46	G3A15	84	G4B01	90
G1A16	30			G2D11	150	G3A16	83	G4B02	91
		G2 – Operating		G2D12	29	G3A17	85	G4B03	104
G1B01	171	**Procedures**				G3A18	80	G4B04	162
G1B02	56	G2A01	50	G2E01	59	G3A19	85	G4B05	91
G1B03	55	G2A02	51	G2E02	60			G4B06	91
G1B04	68	G2A03	50	G2E03	62	G3B01	73	G4B07	122
G1B05	37	G2A04	50	G2E04	57	G3B02	73	G4B08	163
G1B06	37	G2A05	49	G2E05	58	G3B03	72	G4B09	97
G1B07	38	G2A06	49	G2E06	58	G3B04	73	G4B10	163
G1B08	36	G2A07	50	G2E07	57	G3B05	72	G4B11	164
G1B09	38	G2A08	49	G2E08	63	G3B06	76	G4B12	162
G1B10	55	G2A09	51	G2E09	59	G3B07	76	G4B13	162
G1B11	35	G2A10	47	G2E10	60	G3B08	72	G4B14	163
G1B12	36	G2A11	47	G2E11	60	G3B09	71	G4B15	167
G1B13	36	G2A12	45			G3B10	74	G4B16	123
		G2A13	45	G2F01	55	G3B11	76		
G1C01	28			G2F02	53	G3B12	74	G4C01	142
G1C02	27	G2B01	46	G2F03	54	G3B13	79	G4C02	142
G1C03	28	G2B02	46	G2F04	55	G3B14	74	G4C03	143
G1C04	27	G2B03	46	G2F05	52			G4C04	143
G1C05	26	G2B04	52	G2F06	52	G3C01	75	G4C05	140
G1C06	31	G2B05	46	G2F07	52	G3C02	71	G4C06	140
G1C07	29	G2B06	58	G2F08	54	G3C03	71	G4C07	141
G1C08	61	G2B07	32	G2F09	54	G3C04	71	G4C08	142
G1C09	62	G2B08	32	G2F10	54	G3C05	76	G4C09	141
G1C10	61	G2B09	64	G2F11	54	G3C06	77	G4C10	178
G1C11	62	G2B10	59			G3C07	77	G4C11	143
G1C12	58	G2B11	60			G3C08	78	G4C12	144
G1C13	29	G2B12	45			G3C09	77	G4C13	141
		G2B13	52			G3C10	78		
G1D01	24					G3C11	74	G4D01	88
G1D02	41	G2C01	65			G3C12	75	G4D02	88
G1D03	24	G2C02	68			G3C13	148	G4D03	89
G1D04	42	G2C03	69			G3C14	145	G4D04	104
G1D05	41	G2C04	65					G4D05	97
G1D06	23	G2C05	68					G4D06	104
G1D07	43	G2C06	deleted					G4D07	170

Glossary

Amateur communication: Noncommercial radio communication by or among amateur stations solely with a personal aim and without personal or business interest.

Amateur operator/primary station license: An instrument of authorization issued by the FCC comprised of a station license, and also incorporating an operator license indicating the class of privileges.

Amateur operator: A person holding a valid license to operate an amateur station issued by the FCC. Amateur operators are frequently referred to as ham operators.

Amateur Radio services: The amateur service, the amateur-satellite service, and the radio amateur civil emergency service.

Amateur-satellite service: A radiocommunication service using stations on Earth satellites for the same purpose as those of the amateur service.

Amateur service: A radiocommunication service for the purpose of self-training, intercommunication and technical investigations carried out by amateurs; that is, duly authorized persons interested in radio technique solely with a personal aim and without pecuniary interest.

Amateur station: A station licensed in the amateur service embracing necessary apparatus at a particular location used for amateur communication.

AMSAT: Radio Amateur Satellite Corporation, a nonprofit scientific organization. (P.O. Box #27, Washington, DC 20044)

ANSI: American National Standards Institute. A non-government organization that develops recommended standards for a variety of applications.

APRS: Automatic Position Radio System, which takes GPS (Global Positioning System) information and translates it into an automatic packet of digital information.

ARES: Amateur Radio Emergency Service — the emergency division of the American Radio Relay League. Also see RACES

ARRL: American Radio Relay League, national organization of U.S. Amateur Radio operators. (225 Main Street, Newington, CT 06111)

Audio Frequency (AF): The range of frequencies that can be heard by the human ear, generally 20 hertz to 20 kilohertz.

Automatic control: The use of devices and procedures for station control without the control operator being present at the control point when the station is transmitting.

Automatic Volume Control (AVC): A circuit that continually maintains a constant audio output volume in spite of deviations in input signal strength.

Beam or Yagi antenna: An antenna array that receives or transmits RF energy in a particular direction. Usually rotatable.

Block diagram: A simplified outline of an electronic system where circuits or components are shown as boxes.

Broadcasting: Information or programming transmitted by radio means intended for the general public.

Bulletin No. 65: The Office of Engineering & Technology bulletin that provides specified safety guidelines for human exposure to radiofrequency (RF) radiation.

Business communications: Any transmission or communication the purpose of which is to facilitate the regular business or commercial affairs of any party. Business communications are prohibited in the amateur service.

Call Book: A published list of all licensed amateur operators available in North American and Foreign editions.

Call sign: The FCC systematically assigns each amateur station its primary call sign.

Certificate of Successful Completion of Examination (CSCE): A certificate providing examination credit for 365 days. Both written and code credit can be authorized.

Coaxial cable, Coax: A concentric, two-conductor cable in which one conductor surrounds the other, separated by an insulator.

Controlled Environment: Involves people who are aware of and who can exercise control over radiofrequency exposure. Controlled exposure limits apply to both occupational workers and Amateur Radio operators and their immediate households.

Control operator: An amateur operator designated by the licensee of an amateur station to be responsible for the station transmissions.

Coordinated repeater station: An amateur repeater station for which the transmitting and receiving frequencies have been recommended by the recognized repeater coordinator.

Coordinated Universal Time (UTC): (Also Greenwich Mean Time, UCT or Zulu time.) The time at the zero-degree (0°) Meridian which passes through Greenwich, England. A universal time among all amateur operators.

Crystal: A quartz or similar material which has been ground to produce natural vibrations of a specific frequency. Quartz crystals produce a high degree of frequency stability in radio transmitters.

CW: See Morse code.

Dipole antenna: The most common wire antenna. Length is equal to one-half of the wavelength. Fed by coaxial cable.

Dummy antenna: A device or resistor which serves as a transmitter's antenna without radiating radio waves. Generally used to tune up a radio transmitter.

Duplexer: A device that allows a single antenna to be simultaneously used for both reception and transmission.

Duty cycle: As applies to RF safety, the percentage of time that a transmitter is "on" versus "off" in a 6- or 30-minute time period.

Effective Radiated Power (ERP): The product of the transmitter (peak envelope) power, expressed in watts, delivered to the antenna, and the relative gain of an antenna over that of a half-wave dipole antenna.

Electromagnetic radiation: The propagation of radiant energy, including infrared, visible light, ultraviolet, radiofrequency, gamma and X-rays, through space and matter.

Emergency communication: Any amateur communication directly relating to the immediate safety of life of individuals or the immediate protection of property.

Examination Element: The written theory exam or CW test required for various classes of FCC Amateur Radio licenses. Technician must pass Element 2 written theory; General must pass Element 3 written theory plus Element 1 CW; Extra must pass Element 4 written theory.

Far Field: The electromagnetic field located at a great distance from a transmitting antenna. The far field begins at a distance that depends on many factors, including the wavelength and the size of the antenna. Radio signals are normally received in the far field.

FCC Form 605: The FCC application form used to apply for a new amateur operator/primary station license or to renew or modify an existing license.

Federal Communications Commission (FCC): A board of five Commissioners, appointed by the President, having the power to regulate wire and radio telecommunications in the U.S.

Feedline: A system of conductors that connects an antenna to a receiver or transmitter.

Field Day: Annual activity sponsored by the ARRL to demonstrate emergency preparedness of amateur operators.

Field strength: A measure of the intensity of an electric or magnetic field. Electric fields are measured in volts per meter; magnetic fields in amperes per meter.

Filter: A device used to block or reduce alternating currents or signals at certain frequencies while allowing others to pass unimpeded.

Frequency: The number of cycles of alternating current in one second.

Frequency coordinator: An individual or organization which recommends frequencies and other operating and/or technical parameters for amateur repeater operation in order to avoid or minimize potential interferences.

Frequency Modulation (FM): A method of varying a radio carrier wave by causing its frequency to vary in accordance with the information to be conveyed.

Frequency privileges: The transmitting frequency bands available to the various classes of amateur operators. The various Class privileges are listed in Part 97.301 of the FCC rules.

Ground: A connection, accidental or intentional, between a device or circuit and the earth or some common body and the earth or some common body serving as the earth.

Ground wave: A radio wave that is propagated near or at the earth's surface.

Handi-Ham system: Amateur organization dedicated to assisting handicapped amateur operators. (3915 Golden Valley Road, Golden Valley, MN 55422)

Harmful interference: Interference which seriously degrades, obstructs or repeatedly interrupts the operation of a radio communication service.

Harmonic: A radio wave that is a multiple of the fundamental frequency. The second harmonic is twice the fundamental frequency, the third harmonic, three times, etc.

Hertz: One complete alternating cycle per second. Named after Heinrich R. Hertz, a German physicist. The number of hertz is the frequency of the audio or radio wave.

High Frequency (HF): The band of frequencies that lie between 3 and 30 Megahertz. It is from these frequencies that radio waves are returned to earth from the ionosphere.

High-Pass filter: A device that allows passage of high frequency signals but attenuates the lower frequencies. When installed on a television set, a high-pass filter allows TV frequencies to pass while blocking lower-frequency amateur signals.

Inverse Square Law: The physical principle by which power density decreases as you get further away from a transmitting antenna. RF power density decreases by the inverse square of the distance.

Ionization: The process of adding or stripping away electrons from atoms or molecules. Ionization occurs when substances are heated at high temperatures or exposed to high voltages. It can lead to significant genetic damage in biological tissue.

Ionosphere: Outer limits of atmosphere from which HF amateur communications signals are returned to earth.

IRC: International Reply Coupon, a method of prepaying postage for a foreign amateur's QSL card.

Jamming: The intentional, malicious interference with another radio signal.

Key clicks, Chirps: Defective keying of a telegraphy signal sounding like tapping or high varying pitches.

Linear amplifier: A device that accurately reproduces a radio wave in magnified form.

Long wire: A horizontal wire antenna that is one wavelength or longer in length.

Low-Pass filter: Device connected to worldwide transmitters that inhibits passage of higher frequencies that cause television interference but does not affect amateur transmissions.

Machine: A ham slang word for an automatic repeater station.

Malicious interference: See jamming.

MARS: The Military Affiliate Radio System. An organization that coordinates the activities of amateur communications with military radio communications.

Maximum authorized transmitting power: Amateur stations must use no more than the maximum transmitter power necessary to carry out the desired communications. The maximum P.E.P. output power levels authorized Novices are 200 watts in the 80-, 40-, 15- and 10-meter bands, 25 watts in the 222-MHz band, and 5 watts in the 1270-MHz bands.

Maximum Permissible Exposure (MPE): The maximum amount of electric and magnetic RF energy to which a person may safely be exposed.

Maximum usable frequency (MFU): The highest frequency that will be returned to earth from the ionosphere.

Medium frequency (MF): The band of frequencies that lies between 300 and 3,000 kHz (3 MHz).

Microwave: Electromagnetic waves with a frequency of 300 MHz to 300 GHz. Microwaves can cause heating of biological tissue.

Mobile operation: Radio communications conducted while in motion or during halts at unspecified locations.

Mode: Type of transmission such as voice, teletype, code, television, facsimile.

Modulate: To vary the amplitude, frequency, or phase of a radiofrequency wave in accordance with the information to be conveyed.

Morse code: The International Morse code, A1A emission. Interrupted continuous wave communications conducted using a dot-dash code for letters, numbers and operating procedure signs.

Near Field: The electromagnetic field located in the immediate vicinity of the antenna. Energy in the near field depends on the size of the antenna, its wavelength and transmission power.

Nonionizing radiation: Electromagnetic waves, or fields, which do not have the capability to alter the molecular structure of substances. RF energy is nonionizing radiation.

Novice operator: An FCC licensed, entry-level amateur operator in the amateur service.

Occupational exposure: See controlled environment.

OET: Office of Engineering & Technology, a branch of the FCC that has developed the guidelines for radiofrequency (RF) safety.

Ohm's law: The basic electrical law explaining the relationship between voltage, current and resistance. The current (I) in a circuit is equal to the voltage (E) divided by the resistance (R), or $I = E/R$.

OSCAR: "Orbiting Satellite Carrying Amateur Radio." A series of satellites designed and built by amateur operators of several nations.

Oscillator: A device for generating oscillations or vibrations of an audio or radiofrequency signal.

Packet radio: A digital method of communicating computer-to-computer. A terminal-node controller makes up the packet of data and directs it to another packet station.

Peak Envelope Power (PEP): 1. The power during one radiofrequency cycle at the crest of the modulation envelope, taken under normal operating conditions. 2. The maximum power that can be obtained from a transmitter.

Phone patch: Interconnection of amateur radio to the public switched telephone network, and operated by the control operator of the station.

Power density: A measure of the strength of an electro-magnetic field at a distance from its source. Usually expressed in milliwatts per square centimeter (mW/cm2). Far-field power density decreases according to the Law of Inverse Squares.

Power supply: A device or circuit that provides the appropriate voltage and current to another device or circuit.

Propagation: The travel of electromagnetic waves or sound waves through a medium.

Public exposure: See "uncontrolled" environment.

Q-signals: International three-letter abbreviations beginning with the letter Q used primarily to convey information using the Morse code.

QSL Bureau: An office that bulk processes QSL (radio confirmation) cards for (or from) foreign amateur operators as a postage-saving mechanism.

RACES (Radio Amateur Civil Emergency Service): A radio service using amateur stations for civil defense communications during periods of local, regional, or national emergencies.

Radiation: Electromagnetic energy, such as radio waves, traveling forth into space from a transmitter.

Radiofrequency (RF): The range of frequencies over 20 kilohertz that can be propagated through space.

Radiofrequency (RF) radiation: Electromagnetic fields or waves having a frequency between 3 kHz and 300 GHz.

Radiofrequency spectrum: The eight electromagnetic bands ranked according to their frequency and wavelength. Specifically, the very-low, low, medium, high, very-high, ultra-high, super-high, and extremely-high frequency bands.

Radio wave: A combination of electric and magnetic fields varying at a radiofrequency and traveling through space at the speed of light.

Repeater operation: Automatic amateur stations that retransmit the signals of other amateur stations.

Routine RF radiation evaluation: The process of determining if the RF energy from a transmitter exceeds the Maximum Permissible Exposure (MPE) limits in a controlled or uncontrolled environment.

RST Report: A telegraphy signal report system of Readability, Strength and Tone.

S-meter: A voltmeter calibrated from 0 to 9 that indicates the relative signal strength of an incoming signal at a radio receiver.

Selectivity: The ability of a circuit (or radio receiver) to separate the desired signal from those not wanted.

Sensitivity: The ability of a circuit (or radio receiver) to detect a specified input signal.

Short circuit: An unintended, low-resistance connection across a voltage source resulting in high current and possible damage.

Shortwave: The high frequencies that lie between 3 and 30 Megahertz that are propagated long distances.

Single-Sideband (SSB): A method of radio transmission in which the RF carrier and one of the sidebands is suppressed and all of the information is carried in the one remaining sideband.

Skip wave, Skip zone: A radio wave reflected back to earth. The distance between the radio transmitter and the site of a radio wave's return to earth.

Sky-wave: A radio wave that is refracted back to earth. Sometimes called an ionospheric wave.

Specific Absorption Rate (SAR): The time rate at which radiofrequency energy is absorbed into the human body.

Spectrum: A series of radiated energies arranged in order of wavelength. The radio spectrum extends from 20 kilohertz upward.

Spurious Emissions: Unwanted radiofrequency signals emitted from a transmitter that sometimes cause interference.

Station license, location: No transmitting station shall be operated in the amateur service without being licensed by the FCC. Each amateur station shall have one land location, the address of which appears in the station license.

Sunspot Cycle: An 11-year cycle of solar disturbances which greatly affects radio wave propagation.

Technician operator: An Amateur Radio operator who has successfully passed Element 2.

Technician-Plus: An amateur operator who has passed a 5-wpm code test in addition to Technician Class requirements.

Telegraphy: Communications transmission and reception using CW, International Morse code.

Telephony: Communications transmission and reception in the voice mode.

Telecommunications: The electrical conversion, switching, transmission and control of audio video and data signals by wire or radio.

Temporary operating authority: Authority to operate your amateur station while awaiting arrival of an upgraded license.

Terrestrial station location: Any location of a radio station on the surface of the earth including the sea.

Thermal effects: As applies to RF radiation, biological tissue damage resulting because of the body's inability to cope with or dissipate excessive heat.

Third-party traffic: Amateur communication by or under the supervision of the control operator at an amateur station to another amateur station on behalf of others.

Time-averaging: As applies to RF safety, the amount of electromagnetic radiation over a given time. The premise of time-averaging is that the human body can tolerate the thermal load caused by high, localized RF exposures for short periods of time.

Transceiver: A combination radio transmitter and receiver.

Transition region: Area where power density decreases inversely with distance from the antenna.

Transmatch: An antenna tuner used to match the impedance of the transmitter output to the transmission line of an antenna.

Transmitter: Equipment used to generate radio waves. Most commonly, this radio carrier signal is amplitude varied or frequency varied (modulated) with information and radiated into space.

Transmitter power: The average peak envelope power (output) present at the antenna terminals of the transmitter. The term "transmitted" includes any external radio-frequency power amplifier which may be used.

Ultra High Frequency (UHF): Ultra high frequency radio waves that are in the range of 300 to 3,000 MHz.

Uncontrolled environment: Applies to those persons who have no control over their exposure to RF energy in the environment. Residences adjacent to ham radio installations are considered to be in an "uncontrolled" environment.

Upper Sideband (USB): The proper operating mode for sideband transmissions made in the new Novice 10-meter voice band. Amateurs generally operate USB at 20 meters and higher frequencies; lower sideband (LSB) at 40 meters and lower frequencies.

Very High Frequency (VHF): Very high frequency radio waves that are in the range of 30 to 300 MHz.

Volunteer Examiner: An amateur operator of at least a General Class level who prepares and administers amateur operator license examinations.

Volunteer Examiner Coordinator (VEC): A member of an organization which has entered into an agreement with the FCC to coordinate the efforts of volunteer examiners in preparing and administering examinations for amateur operator licenses.

Index

A special invitation from Gordon West, WB6NOA, ARRL Life Member

"Join ARRL and experience the BEST of Ham Radio!"

ARRL Membership Benefits and Services:

- *QST* magazine — your monthly source of news, easy-to-read product reviews, and features for new hams!
- Technical Information Service — access to problem-solving experts!
- Members-only Web services — find information fast, anytime!
- ARRL clubs, mentors and volunteers — ready to welcome YOU!

FREE Book Offer!

Getting Started with **HAM RADIO**

A Guide to your FIRST Amateur Radio Station

I want to join ARRL.

Send me the FREE book I have selected (choose one)

- ☐ **Getting Started with Ham Radio**
- ☐ **The ARRL Repeater Directory**
- ☐ **More Wire Antenna Classics**

Explore the World of HF Radio

Name _____ Call Sign _____

Street _____

City _____ State _____ ZIP _____

Email _____

Please check the appropriate one-year¹ rate:
- ☐ **$39 in US.**
- ☐ **Age 65 or older rate, $36 in US.**
- ☐ **Age 21 or younger rate, $20 in US** (see note*).
- ☐ **Canada $49.**
- ☐ **Elsewhere $62.**
- **Please indicate date of birth** _____ .

¹ 1-year membership dues include $15 for a 1-year subscription to QST. International 1-year rates include a $10 surcharge for surface delivery to Canada and a $23 surcharge for air delivery to other countries. Other US membership options available: Blind, Life, and QST by First Class postage. Contact ARRL for details.
*Age 21 or younger rate applies only if you are the oldest licensed amateur in your household.
International membership is available with an annual CD-ROM option (no monthly receipt of QST). Contact ARRL for details.
Dues subject to change without notice.

Sign up my family members, residing at the same address, as ARRL members too! They'll each pay only $8 for a year's membership, have access to ARRL benefits and services (except QST) and also receive a membership card.

☐ Sign up _____ family members @ $8 each = $ _____ . Attach their names & call signs (if any).

☐ Total amount enclosed, payable to ARRL $ _____ . (US funds drawn on a bank in the US).

☐ Enclosed is $ _____ ($1.00 minimum) as a donation to the Legal Research and Resource Fund.

☐ Charge to: ☐ VISA   ☐ MasterCard   ☐ Amex   ☐ Discover

Card Number _____ Expiration Date _____

Cardholder's Signature _____

Call Toll-Free (US) **1-888-277-5289**
Join Online **www.arrl.org/join** or
Clip and send to:

☐ If you do not want your name and address made available for non-ARRL related mailings, please check here.

ARRL *The national association for* **AMATEUR RADIO**
225 Main Street
Newington, CT 06111-1494 USA

Web Code: GWX

FREE
CQ Mini-Sub!

We'd like to introduce you to a Ham radio magazine that's fun to read, interesting from cover to cover and written so that you can understand it—**FREE!** The magazine is **CQ Amateur Radio**—read by thousands of people every month in 116 countries around the world. Get your FREE 3 issue mini-sub, courtesy **CQ** and **Master Publishing**.

CQ is aimed squarely at the **active** ham. You'll find features and columns covering the broad and varied landscape of the amateur radio hobby from contesting and DXing to satellites and the latest digital modes. Equipment reviews, projects and articles on the science as well as the art of radio communications—all in the pages of **CQ Amateur Radio**.

Reserve your FREE 3-month mini-sub today!

Remove this page, fill in your information below, fold, tape closed with the postage paid **CQ** address showing and mail today!

Send my FREE 3-issue CQ Mini-Sub to:

Name _____

Address _____

City _____ State _____ Zip _____

Make a great deal even better!

Add a one-year CQ subscription to your FREE mini-sub at a
**Special Introductory Rate—15 issues for only $29.95
a 60% savings off the newsstand price!**

Yes! I want to take advantage of this special 15-issue offer.

☐ Check/Money Order enclosed.
Bill my: ☐ Visa ☐ MasterCard ☐ AMEX ☐ Discover
Insert your credit card number below:

Credit Card Expiration Date: _____

Fax your order to: 516-681-2926
Visit our web site: www.cq-amateur-radio.com

YI-G-07

CUT HERE ✂

TAPE

FREE
CQ Amateur Radio Mini-Sub!

Fill in your address information
on the reverse side of this page, fold in half,
tape closed and mail today!

TAPE

TAPE

✂ CUT HERE

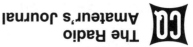

Hicksville, NY 11801-9962
25 Newbridge Road

The Radio
Amateur's Journal

POSTAGE WILL BE PAID BY ADDRESSEE

FIRST CLASS MAIL PERMIT NO. 2055 HICKSVILLE, NY 11801

BUSINESS REPLY MAIL

NO POSTAGE
NECESSARY
IF MAILED
IN THE
UNITED STATES